Blockchain for Industry 4.0

This reference text provides the theoretical foundations, the emergence, and the application areas of blockchain in an easy-to-understand manner that would be highly helpful for researchers, academicians, and industry professionals to understand the disruptive potential of blockchain. It explains blockchain concepts related to Industry 4.0, smart healthcare, and Internet of Things (IoT) and explores smart contracts and consensus algorithms. This book will serve as an ideal reference text for graduate students and academic researchers in electrical engineering, electronics and communication engineering, computer engineering, and information technology.

This book

- Discusses applications of blockchain technology in diverse sectors such as Industry 4.0, education, finance, and supply chain.
- Provides theoretical concepts, applications, and research advancements in the field of blockchain.
- Covers the Industry 4.0 digitization platform and blockchain for data management in Industry 4.0 in a comprehensive manner.
- Emphasizes analysis and design of consensus algorithms, fault tolerance, and strategies to choose the correct consensus algorithm.
- Introduces security issues in the industrial IoT, IoT, blockchain integration, and blockchain-based applications.

The text presents in-depth coverage of theoretical concepts, applications, and advances in the field of blockchain technology. This book will be an ideal reference for graduate students and academic researchers in diverse engineering fields such as electrical, electronics and communication, computer, and information technology.

Computational Methods for Industrial Applications

Series Editor:
Bharat Bhushan

In today's world, IoT platforms and processes in conjunction with the disruptive blockchain technology and path-breaking artificial intelligence (AI) algorithms lay out a sparking and stimulating foundation for sustaining smarter systems. Further computational intelligence (CI) has gained enormous interest from various quarters in order to solve numerous real-world problems and enable intelligent behavior in changing and complex environments. This book series focuses on varied computational methods incorporated within the system with the help of AI, learning methods, analytical reasoning, and sense making in big data. Aimed at graduate students, academic researchers, and professionals, the proposed series will cover the most efficient and innovative technological solutions for industrial applications and sustainable smart societies in order to alter green power management, effects of carbon emissions, air quality metrics, industrial pollution levels, biodiversity, and ecology.

Blockchain for Industry 4.0: Emergence, Challenges, and Opportunities
Anoop V. S., Asharaf S., Justin Goldston, and Samson Williams

Intelligent Systems and Machine Learning for Industry: Advancements, Challenges, and Practices
P.R Anisha, C. Kishor Kumar Reddy, Nguyen Gia Nhu, Megha Bhushan, Ashok Kumar, and Marlia Mohd Hanafiah

Blockchain for Industry 4.0

Emergence, Challenges, and Opportunities

Edited by
Anoop V. S., Asharaf S., Justin Goldston, and
Samson Williams

CRC Press
Taylor & Francis Group
Boca Raton London New York

CRC Press is an imprint of the
Taylor & Francis Group, an **informa** business

First edition published 2023
by CRC Press
6000 Broken Sound Parkway NW, Suite 300, Boca Raton, FL 33487-2742

and by CRC Press
4 Park Square, Milton Park, Abingdon, Oxon, OX14 4RN

CRC Press is an imprint of Taylor & Francis Group, LLC

Library of Congress CataloginginPublication Data
Names: S., Anoop V., editor.
Title: Blockchain for industry 4.0 : emergence, challenges, and opportunities / edited by Anoop V. S., Asharaf S., Justin Goldston and Samson Williams.
Description: First edition. | Boca Raton, FL : CRC Press, 2023. | Series: Computational methods for industrial applications | Includes bibliographical references and index.
Identifiers: LCCN 2022030561 (print) | LCCN 2022030562 (ebook) | ISBN 9781032253664 (hbk) | ISBN 9781032253695 (pbk) | ISBN 9781003282914 (ebk)
Subjects: LCSH: Blockchains (Databases)–Industrial applications. | Industry 4.0.
Classification: LCC QA76.9.B56 B553 2023 (print) | LCC QA76.9.B56 (ebook) | DDC 005.74-dc23/eng/20220830
LC record available at https://lccn.loc.gov/2022030561
LC ebook record available at https://lccn.loc.gov/2022030562

ISBN: 9781032253664 (hbk)
ISBN: 9781032253695 (pbk)
ISBN: 9781003282914 (ebk)

DOI: 10.1201/9781003282914

Typeset in Sabon
by codeMantra

Contents

Preface

Blockchain, a distributed ledger management infrastructure where blocks of transactional data are hash chained for immutability to establish trust for enabling the digital currency Bitcoin, has recently emerged as a disruptive technology with many real-world applications demanding trusted data management. Such a blockchain infrastructure on clusters of computers providing decentralized data management is now believed to have the capability to move the notion of trust from trusted institutional mechanisms to mathematically provable technology infrastructure. Since trust in data is the key to the management for establishing relationships between businesses, government, and people, blockchain can definitely bring disruptions in a vast array of domains and sectors. Due to this disruptive nature, blockchain technology is getting a great deal of attention from industries and businesses which are vying with each other to incorporate this immutable, shared, and tamper-proof ledger mechanism into their business models. There are widespread opinions that claim blockchain just as hype, but very recently, we have witnessed the business adoption graphs going high. Some of the industry and business giants have already implemented this distributed ledger technology in their mainstream business modules to ensure the credibility of transactions for their stakeholders. The scope of this proposed edited book is to discuss the theoretical foundations, the emergence, and the application areas of blockchain in an easy-to-understand manner that would be highly helpful for researchers, academicians, and industry professionals to understand the disruptive potentials of blockchain. This book provides the theoretical concepts, empirical studies, and detailed overview of various aspects related to the development of blockchain applications from a reliable, trusted, and secure data transmission perspective. Further, it also intends to induce further research about the ethical impact of the new "distributed trust" paradigm resulting from the surge of such a disruptive technology.

Editors

Anoop V. S. is a Postdoctoral Fellow at Smith School of Business, Queen's University, Ontario, Canada. Previously, he worked as a Senior Scientist (Research and Training) at Kerala Blockchain Academy, Kerala University of Digital Sciences, Innovation, and Technology, Thiruvananthapuram, India. He received his Doctor of Philosophy (PhD) and Master of Philosophy (MPhil) from Cochin University of Science and Technology (CUSAT), Kerala. He has also worked as a Senior AI Data Scientist with Dubai Future Accelerator for implementing AI solutions for Etisalat and Etihad Airways. He is an experienced software engineer with more than 5 years of experience working with US-based multi-national companies in Technopark, Thiruvananthapuram. He has completed several industry internships in data science and AI that helped him implement intelligent algorithms for many organizations to solve their complex business problems. He has over 10 years of overall experience in teaching, research, and industry. His research interests are in text mining, natural language processing, computational social sciences, and blockchain. He has published papers in journals of national and international repute, book chapters, and conference proceedings in major venues of AI and knowledge management.

Asharaf S. is a computer engineer/scientist with extensive exposure in algorithms for machine learning. In addition to his current role as a Dean (R&D) and Professor at Kerala University of Digital Sciences, Innovation, and Technology, he also serves as a visiting faculty for MDP programs at the Indian Institute of Management Kozhikode and as a Mentor in Kerala Startup Mission. He received his PhD and Master of Engineering degrees in Computer Science from the Indian Institute of Science, Bangalore. He graduated in Computer Engineering from Cochin University of Science and Technology. After B.Tech, he has worked as a lecturer at T.K.M. College of Engineering, Kollam, and after PhD, he has worked as a Research Scientist at America Online (AOL) R&D Labs and as an Assistant Professor at the Indian Institute of Management Kozhikode. He is a recipient of IBM outstanding PhD student award

2006, IBM Shared University Research Grant 2015, IBM Open Science Collaboration Program Grant 2017, and DBT/BIRAC/Bill & Melinda Gates Foundation Research Grant under the Grand Challenges India program in 2019. He has published three books and more than 50 research papers in international journals and conferences. He served as the founding professor-in-charge of Maker Village Cochin (https://makervillage.in/people.php) and currently serves as the Professor-in-Charge of Kerala Blockchain Academy (https://kba.ai/team/). His areas of interest include technologies and business models related to data engineering, machine learning, information retrieval, and blockchains.

Justin Goldston is a Professor of Project and Supply Chain Management at Penn State University, where his research is focused on blending the practices of supply chain management, emerging technologies, and sustainability to create positive global change. Dr. Goldston is a research faculty affiliate for the Center for the Business of Sustainability at the Smeal College of Business at Penn State as well as an active contributing faculty member of The Sustainability Institute at Penn State. Outside of the institution, Goldston is an Executive on the International Supply Chain Education Alliance's (ISCEA's) International Standards Board (IISB) and is the author of the forthcoming book AI for Good: Achieving Sustainability Through Citizen Science and Organizational Citizenship. Dr. Goldston has over 20 years of experience working with organizations around the world such as Intel, Siemens, and Blue Buffalo on business performance improvement, organizational change, and enterprise-wide digital transformation initiatives. Goldston has also led and assisted in the development of Supply Chain Management, Sustainability, and Business Analytics programs and courses at Georgetown University, Texas A&M University, Rasmussen University, Davenport University, and North Carolina Wesleyan College and has evaluated doctoral programs for the Department of Higher Education. Goldston serves on the Management Advisory Board at various higher education institutions; is the author of multiple peer-reviewed journal articles on supply chain management, sustainability, and innovative technologies of Critical Success Factors in ERP Implementations; and is a five-time TEDx speaker where he discussed emerging technologies such as blockchain and AI.

Samson Williams is an Adjunct Professor at Columbia University in NYC and the University of New Hampshire School of Law, where he teaches on the blockchain, cryptocurrencies, and the space economy. He is also a serial entrepreneur and accidental investor and the President of the Crowdfunding Professional Association and investor into two investment crowdfunding platforms Brite.us – CrowdInvesting Done B_rite and GoingPublic.com. Samson is a senior emergency and finance strategist specializing in operations and technology, healthcare, and financial service crisis management. As a globally recognized expert on technology and crisis communications, Samsons strive to develop compelling content and deliver a message through distinct, actionable, and measurable campaigns. Previously, he worked with Fannie Mae as part of the emergency management team assembled to help pull the nation out of the housing crisis. Ultimately, he spent the last 2 years leading internal communications, workforce development, and organizational resiliency for Fannie Mae's Operations and Technology Division. Prior to investing in Brite.us and creating The Space Economy show, Samson served as the Principal Consultant for Axes and Eggs, a global technology advisory building relationship beyond tech in Dubai and Riyadh and throughout the Middle East North Africa region. There, Mr. Samson oversaw education, messaging, communication strategy, and media planning for finance, fintech, and government agencies throughout the Middle East, executing events and relationships with government agencies from the United Arab Emirates, Kingdoms of Bahrain and Saudi Arabia, Mauritius, and India. Samson has authored several books such as Raising Money – Understanding Cryptocurrencies, Crowdfunding, Startup Capital, Blockchain and The Space Economy, The Space Economy – Book Zero of The Space Economies Series, Lifestyle Money: Blockchain, Cryptocurrencies, Crowdfunding, and The Future of Money and Wealth, Stop Sucking Up to VCs, and Race and Space – Racial Equity and Justice in the Space Economy.

Contributors

Shugufta Abrahim Wani
Graduate School of Science and Engineering
University of Toyama
Toyama, Japan

Vijay Anant Athavale
Walchand Institute of Technology
Maharashtra, India

Ankit Bansal
Chitkara University
Chandigarh, India

Xavier Boyen
Queensland University of Technology
Brisbane, Australia

Shoufeng Cao
Queensland University of Technology,
Brisbane, Australia

Nikhil V. Chandran
Kerala Blockchain Academy, Kerala University of Digital Sciences,
    Innovation and Technology,
Thiruvananthapuram, India

Felicity Deane
Queensland University of Technology
Brisbane, Australia

Alex Devassy
Ernst and Young GDS
Thiruvananthapuram, India

S. Dhingra
Guru Gobind Singh Indraprastha University
New Delhi, India

Ashutosh Dubey
Research and Innovation Vertical
National Payment Corporation of India
Mumbai, India

Himanshu Falwadiya
Guru Gobind Singh Indraprastha University
New Delhi, India

Marcus Foth
Queensland University of Technology
Brisbane, Australia

Masayoshi Fukushima
Graduate School of Science and Engineering
University of Toyama
Toyama, Japan

Usharani Hareesh Govindarajan
Business School
University of Shanghai for Science and Technology
Shanghai, China

Megha Gupta
Manipal University
Jaipur, India

S. Gupta
Guru Gobind Singh Indraprastha University
New Delhi, India

Shanmugham D. Jayan
Vijayaraghavan and Devi Associates
Kochi, India

Jayanthi M.
CMR University
Bangalore, India

Franklin John
Kerala Blockchain Academy
Kerala University of Digital Sciences,
Innovation and Technology
Thiruvananthapuram, India

Aurangjeb Khan
CMR University
Bangalore, India

Yusera Farooq Khan
Shri Mata Vaishno Devi University
Katra, India

Dhanith Krishna
Ernst and Young GDS
Thiruvananthapuram, India

Zhen Li
Graduate School of Science and Engineering
University of Toyama
Toyama, Japan

Sicheng Liu
Graduate School of Science and Engineering
University of Toyama
Toyama, Japan

Bilal Ahmed Mir
Graduate School of Science and Engineering
University of Toyama
Toyama, Japan

Tanveer A. Mir
Organ Transplant Centre of Excellence
Transplantation Research and Innovation Department
King Faisal Specialist Hospital and Research Centre
Riyadh, Saudi Arabia

Thomas Miller
Queensland University of Technology
Brisbane, Australia

Gagan Narang
Institute of Informatics and Communication
University of Delhi
New Delhi, India

Vidyashankar Ramalingam
Independent Blockchain Consultant
Chennai, India

Renuka V
School of Legal Studies
Cochin University of Science and Technology
Kochi, India

Manju Sadasivan
CMR University
Bangalore, India

Sreelakshmi S.
University of Kerala
Thiruvananthapuram, India

Radha Sridharan
CMR University
Bangalore, India

Swaraj M.
Indian Institute of Management
Kozhikode, India

Anjali Tiwari
Sir HN Reliance Foundation Hospital
Mumbai, India

Deepnarayan Tiwari
Research and Innovation Vertical
National Payment Corporation of India
Mumbai, India

Suhasini Verma
Manipal University
Jaipur, India

Vinod Chandra S. S.
University of Kerala
Thiruvananthapuram, India

Shadil Ibrahim Wani
Graduate School of Science and Engineering
University of Toyama
Toyama, Japan

Vinay Surendra Yadav
National Institute of Technology
Raipur, India

Takuto Yamaguchi
Graduate School of Science and Engineering
University of Toyama
Toyama Japan

Nikola Zivlak
Emlyon Business School
Guy de Collongue
Écully, France

Chapter 1

The rise of blockchain technology in healthcare for the preservation of patient-centric health records

Bilal Ahmed Mir, Sicheng Liu, Zhen Li, and Shugufta Abrahim Wani
University of Toyama

Tanveer A. Mir
King Faisal Specialist Hospital and Research Centre

Yusera Farooq Khan
Shri Mata Vaishno Devi University

Shadil Ibrahim Wani, Masayoshi Fukushima, and Takuto Yamaguchi
University of Toyama

CONTENTS

DOI: 10.1201/9781003282914-1

1

1.1   INTRODUCTION

Blockchain is an emerging scientific and technological domain that is being employed to create innovative solutions in a variety of fields to alleviate the reliance on centralized institutions to certify data integrity and to enable secure electronic data exchange between interacting individuals or organizations (Boek et al., 2017). Blockchain technology holds tremendous potential to revolutionize the biomedical industry and the digital healthcare sector (Jamil et al., 2019). Blockchain-based patient care systems are currently being applied to medical information systems, digital medical information storage and exchange of patient or disease-specific health records, interoperability of patient data and analytics, telemedicine adoption, remote patient monitoring and data acquisition systems, and medical informatics research (Djabarulla et al., 2021). The application of this technology in healthcare is very significant not only for improving the patient care ecosystem and increasing engagement and empowerment but also for enabling healthcare professionals to access information and use the data for proper diagnosis and medical research purposes. Blockchain health information technology (HIT) offers several benefits such as ensuring the patients' privacy, high-security protection of sensitive medical records among healthcare professionals and associated hospitals, diagnostic laboratories, pharmacies, scientists, for diagnosis, treatment, and research (Jamal et al., 2009; Ahram et al., 2017; Dagher et al., 2018). However, the main challenges associated with using blockchain methodology for the production of health information networks include the patients' attitude, perception, and reluctance toward blockchain-linked networks, security risk and privacy concerns, transparency of sharing patient data, high-processing power demands, and implementation and performance analysis expenses, as well as ineffective consensus processes (Ekblaw et al., 2016; Ichikawa et al., 2017; Griggs et al., 2014). However, the majority of blockchain-based health care efforts are still in their early stages (Roehrs et al., 2017; Fan et al., 2018; Zhang et al., 2018). The blockchain has several advantages, including data authenticity, data privacy, data transparency, data scale, and digital assets. The quantity of health data that may be collected has increased due to the digitalization of physical records, sensors, and other technological devices. Data for health care is generated from several sources, including hospital records, radiographic images, and sensors (Raghupathi et al., 2014; Ray et al., 2020). Because it is the patient's personal information, some of the information contained in the different databases belonging to patients is secret. Because of its immutability and data traceability, the use of blockchain technology is an effective strategy for ensuring better security in certain instances. The ability of blockchain technology to monitor the distribution of drugs and ensure that substantial financial losses, drug delivery management, and the right supply chain pattern are all correctly monitored is an important component of blockchain technology. Blockchain offers various benefits over other systems since it is a distributed ledger system with immutability and privacy built-in (Svein et al., 2017; Haq et al., 2018). The usage of patient healthcare records has prompted concerns about patient privacy. Personal information is aggregated in these records. The goal of this research is to look at how blockchain technology is being used in the healthcare business. The study focuses on healthcare data management, as well as medical data exchange, photographs, and reporting. Blockchains have lately received a lot of interest due to their distributed design for value production and administration. Many businesses throughout the globe have deployed or are embracing blockchain technology to improve the integrity, accessibility, and performance of their services.

In contrast to more traditional methods, blockchain technology provides a platform that supports a wide range of direct peer-to-peer transfers of digital assets. Among the categories addressed are cryptocurrency, medical, commerce, banking, copyright infringement, the economy, and societal applications (Nakamoto et al., 2008; Swan et al., 2015; Iansiti et al., 2017). A blockchain is a series of blocks that generate and store encrypted data with digital signatures in a decentralized network while maintaining the privacy, security, transparency, and traceability features of a patient's health record. Researchers in business and academia have begun to study healthcare-related applications using existing blockchain technologies. Despite these advances, there are still issues because blockchain technology has its own set of limits and constraints, such as mining incentives, mining threats, and key management. Even in the face of stringent legislative restraints, such as the Health Insurance Portability and Accountability Act of 1996, blockchain technologies in the healthcare industry frequently need more permission, scalability, and record-sharing agreements (McGhin et al., 2019). It is vital in the digital healthcare era to collect medical data from several sources to aid in data analysis and individualized therapy. The limitations of healthcare organizations' cyber-infrastructure, as well as the possibility of privacy leaks, pose challenges to medical record interchange. Blockchain's decentralization and openness as a public ledger can promote secure medical data exchange. Protected Health Information (PHI) in Electronic Medical Records (EMRs) is particularly susceptible (Anoshiravani et al., 2012). As a result, most healthcare providers and hospitals have strengthened their data security safeguards. As a result, today's medical data firewalls impede collaborative healthcare treatment and medical research. In contrast, cloud computing and big data need the sharing of medical data among many users and organizations to allow analysis and the delivery of enhanced healthcare services and innovative treatment plans.

In this chapter, we aimed to elucidate the applicability of blockchain technology in the healthcare and medical sectors. To assist researchers in this field, we provided an overview of this technology and explored its advantages and limitations. The chapter is organized as follows. In Section 1.2, we introduced the concept of healthcare data and its security and privacy issues. In Section 1.3, we provided an overview of medical data management strategies using blockchain technology. In Section 1.4, we provided an overview of the architecture and taxonomy of blockchain. In Sections 1.5 and 1.6, we highlighted the limitations and conclusions.

1.2 LARGE AMOUNT OF HEALTHCARE DATA AND MEDICAL COMPUTING

Medicine has become an integral part of our lives, and medical data such as patient's medical histories, drug prescriptions, diagnostic test results, and related medical records have become indispensable for patients' diagnosis, medication, recovery progress, and subsequent course of action (Luo et al., 2016). Traditionally, medical records have been preserved on paper, which is readily destroyed or changed. Traditional medical record systems are restricted by a time-consuming administration technique for data processing to preserve patients' privacy, resulting in significant human resource waste. Such a design would stymie medical record exchange.

As a result, it was critical to saving the data in electronic form. Medical databases, on the other hand, maybe tampered with or altogether wiped. There was also the problem of information restriction, which prompted considerable concern. When an entity, such as a person, attempts to get data that should not have been made available to the public without the approval of the data's intended audience, the entity may impede the flow of information.

Technology has always played an important role in resolving numerous challenging obstacles such as quality improvement, proper allocation of resources, information distribution, and storage of information, but data sharing technology for medical care must also evolve with the times. Providing access to medical information while maintaining security and privacy requirements has always been a challenging subject to properly handle. This study includes blockchain-based patient-centric medical information storage and privacy preservation. It was emphasized that the protection of medical information in terms of data security and privacy consists of data confidentiality at the collection and management stage, data integrity during storage, server authentication during the transfer from the local device to cloud or network to network, as well as access control and privacy management. As a result, prioritizing a more pragmatic approach to sharing health data may involve adopting several approaches to accomplish the ultimate goals. Blockchain-based infrastructure and storage platforms represent a new paradigm shift in medical computing and offer several advantages over traditional approaches (Salah et al., 2019). As we observed, the proper type of blockchain must be chosen for medical data transmission. Despite the benefits of blockchain-based medical data management, there are still certain difficulties that must be solved. Our goal is to raise awareness of these challenges by identifying potential research paths and initiatives that improve the security of healthcare data while simultaneously making it simpler to exchange.

1.2.1 Security and privacy of healthcare data

Every organization that handles sensitive information should make it their top responsibility to protect medical data in storage and transit. Cryptographic strategies, including transport layer security (TLS)-based protocols, are considered to be essential cryptographic approaches to secure and protect data in transit. Encryption standards of electronic data, digital signatures, and access control security safety guards are examples of cryptography primitives that can provide secure access to data at rest in a single domain (Liang et al., 2017). On the other hand, implementing multiple cross-domain access control mechanisms to share and protect the exchange of high-value and sensitive data at the state and national levels remains a key challenge. Data privacy and data security are closely related but differ in ensuring the guarantees of lawful collection, use, and protection of personal information (Theodouli et al., 2018). For example, the privacy compliance framework requires that all electronically protected health information (ePHI) operations, including data storage, transfer, and delivery, adhere to standardized security and privacy standards (Ida et al., 2016). Blockchain networks' cryptography guarantees patient confidentiality. Medical data is protected against tampering by data integrity and incorruptibility. Consider the blockchain as a data structure that preserves data in each network node in order to escape the problem of an outage. As a result, there is more consistency, stability, and resistance to assault. The blockchain concept has the potential to resolve the challenging issue related to distributed denial-of-service attacks (DDOS) (Singh et al., 2020). It is possible to speed up the exchange of medical records by introducing blockchain technology into the system. As a result of decentralization, people regain ownership of their medical records, allowing them to manage their health.

The key issue is that healthcare information should be kept safe and confidential from both outsiders and persons who aren't intended to have access to it. There may be a need for novel computing methodologies, such as new architectures or paradigms.

1.3 MEDICAL DATA MANAGEMENT

1.3.1 Health data access and sharing

Blockchain technology has the potential to be utilized to access and share patient medical records (Usman et al., 2020). Patient access to their medical information and history may be difficult owing to the fact that it is dispersed throughout several healthcare facilities. Using MedBlock's distributed blockchain technology, medical records (EMRs) can be accessed and recovered promptly (Shen et al., 2019). Network congestion is prevented because of this software's improved consensus technique. In the case of blockchain, which is designed to develop in complexity over time, there is a potential for network congestion. MedBlock exhibits excellent cryptographic capabilities and access control for healthcare data.

Recently, a blockchain-assisted Medical Data Preservation platform has also been created Data Preservation System (DPS) (Li et al., 2018). Comparable data storage and cryptographic methods are used by DPS to protect data. Blockchain technology is being employed in clinical investigations and, more recently, in medical insurance storage. Blockchains can easily identify fake information by employing smart contracts to safeguard clinical trial operations. These smart contracts are executable codes that can efficiently facilitate the proper recording of tests and provide data that may be utilized to design new cures and drugs. Blockchain can change medical insurance storage as well as clinical trial transformation. Very recently, Zhou reported the development of a blockchain-based medical insurance storage platform (MIStore) that provides high reliability and credibility to interacting users (Shen et al., 2019). Each server in the storage system validates the other server's activity. MIStore can improve the collaboration of patients, caregivers, and insurers.

1.3.2 Need for medical information exchange

Health information interchange is one of the most time-consuming and expensive operations in healthcare. Data privacy is a major concern, particularly in the fields of customized medicine and wearable technologies. So far, data security, sharing, and interoperability have been the most pressing concerns in population health management. The challenge's dependability is ensured through the usage of blockchain technology. In addition to real-time modifications and access, it increases data security, interchange, interoperability, and integrity. Data privacy is a big problem, especially with customized treatment and wearable devices. Blockchain technology enables patients and medical practitioners to securely gather, transmit, and consult data across networks. The healthcare environment is evolving toward a patient-centered strategy that stresses two critical elements: readily available services and enough healthcare resources at all times. This is an area where blockchain might have a huge impact on the healthcare industry. Furthermore, healthcare organizations that use blockchain technology may be able to give their patients superior patient care and healthcare facilities. The problem is solved in a couple of minutes using this strategy. Residents can engage in health research programs by using blockchain technology. Increased research and data exchange on public well-being will also enhance treatment for individuals of various backgrounds and cultures. When a central database is properly established, it is used to regulate changes and access. Because patients and medical workers seek a safe and simple way of recording, transmitting, and consulting data through networks, blockchain technology is being utilized to address these issues in the healthcare sector.

1.3.3 Cloud and cryptography in healthcare

As cloud computing has grown in popularity, it has become increasingly challenging to keep data safe while being shared across several locations. A major problem of cloud data interchange is assessing the amount of trust that customers may place in cloud service providers since they operate in different administrative or security domains. When it comes to the cloud, users are losing control over their data and, as a result, confidence is eroding because of this. It is necessary to shift physical control of data from one trusted network to another, or from local storage, domain to cloud when it is outsourced (cloud storage) (Shen et al., 2019). Users' personal information is stored in a number of locations, both online and in the actual world. Users have no idea where their information is stored or if the security measures in place on these websites are up to par with what they expect.

Cloud-based medical information management has many of the same challenges. In addition, the security and privacy requirements imposed by Health Insurance Portability and Accountability Act (HIPAA) make inter-institutional medical data interchange even more challenging (Rodrigues et al., 2018).

1.3.4 Blockchain-based patient-centric medical information storage

In the past decade, the healthcare record management system has undergone significant changes in order to better serve patients by increasing accuracy, efficiency, and quality. Health records are currently kept on cloud servers and maintained by a third-party cloud service provider under either a client/server model where each institution maintains its database or under a cloud model (Dubovitskaya et al., 2020) because of security, privacy, data leakage, and fragmentation difficulties with these systems. The medical history of a patient from all of their medical care facilities cannot be viewed in one place by healthcare providers or patients (Figure 1.1).

Patient Centricity: In the recent decade, the term "patient centricity" has taken on a new meaning. Patients nowadays are well-informed about their health conditions and want open and honest care (Jabarulla et al., 2021). People may now easily obtain health-related information thanks to smartphones and the internet. To bridge the gap between patients' requirements and current healthcare solutions, healthcare service providers, pharmaceutical companies, and medical practitioners are emphasizing patient centricity. However, the general public must be informed about patient centricity, its importance, and its potential consequences.

Patient-centric technology enables patients to use technology and navigate their clinical information across various institutions via a single login and secure website (Naresh et al., 2021). The system secures patient information and provides security in accordance with the criteria for patient care data framework and access control guidelines. Several of the smart contracts (Mani et al., 2021) include the following:

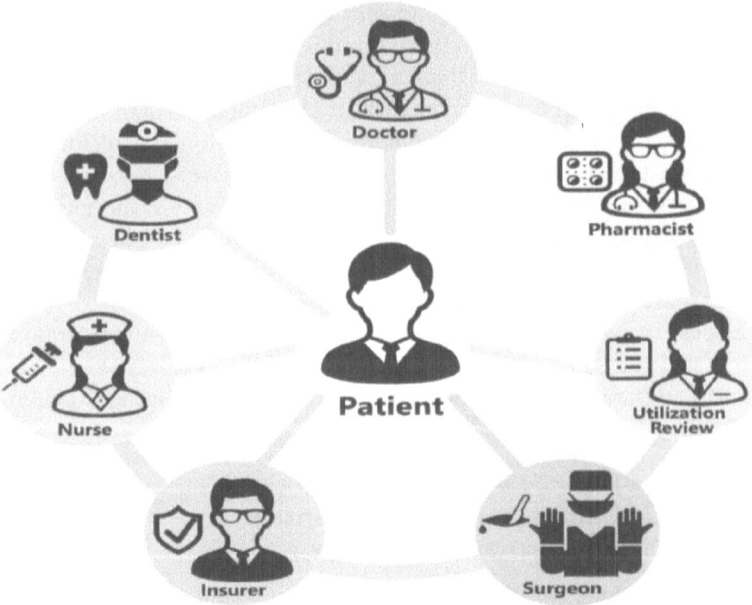

Figure 1.1 The concept of the development of blockchain-based patient-centric health informatics services.

i. Registration agreement for the construction of an immutable patient log system.
ii. Health Record Creation agreement for assistance in generating electronic health data files.
iii. Health Record Storage agreement using a novel Modified Merkle Tree database for safe storage and quick access.
iv. Maintain an up-to-date permission contract that allows for emergency access.
v. Data sharing permission agreement for the exchange of health records between various interacting parties.
vi. Viewership permission agreement allows patients to examine their health information for home care and future treatment.

1.3.5 Major challenges of cloud computing in healthcare management

Healthcare organizations typically deploy information technology infrastructure based on a private cloud design. However, the traditional approaches suffer from poor system performance, low scalability, and limitations related to data sharing. A private cloud is possible when IT services are delivered across a private network to any specific organization (Ekonomou et al., 2011). Because building a scalable private cloud requires a high degree of compatibility and huge investments in computing infrastructure and storage devices, and given the rapidly changing volume of clinical data, it is difficult to accurately predict future cloud capacity requirements. In a private cloud-based strategy, it is difficult for parties outside the domain boundary to access data stored within the domain perimeter. Because of the above-mentioned limitations, medical information cannot be shared in the manner required by big data analytics techniques. Public clouds, on the other hand, are ideal for data scalability and sharing. In public cloud services, multi-tenancy refers to the sharing of virtual machines among many applications, exposing data to a range of assaults. Encryption is a critical necessity for healthcare management in private or public clouds to protect data security and privacy. An undesirable scenario is created by the major management issue. While it is desired to increase data security and provide cloud users greater autonomy, sending encryption keys to authorized users is a time-consuming operation that restricts the scalability of cloud data sharing among different businesses. This was a fundamental problem in previous systems that relied on strategic distribution locations Key Distribution Centers (KDCs) (Bharathy et al., 2014). It allows cloud administrators to "touch" the secret keys and further decrypt the stored data, which significantly raises the danger of data leakage. Externally hosted clouds that manage medical information, such as Amazon, Google, and Microsoft, must solve this quandary owing to HIPAA compliance (Wu et al., 2012).

1.3.6 Blockchain-based healthcare

With the growing usage of blockchain technology as a general trend in distributed computing systems, numerous academics are currently contemplating using blockchain to enhance security standards for sharing and managing medical data to improve patient safety. Technology has always played an important role in improving the overall quality of life and solving various problems such as resource allocation, information transmission, and blockages, and data sharing technology here in medical care must also evolve with the times. Individuals may have a number of medical healthcare service providers, such as primary care physicians, specialists, and even therapists. Because a sickness may be the outcome of a previous ailment, everyone's health information must be shared safely and without change. If the data is carefully stored and communicated, patients can accurately recall all the information, even if they are not professionals or do not have a good memory.

Patients are responsible for keeping their medical records up to date. The rules and regulations for health and liberty vary from country to country. For example, Article 21 of the Indian Constitution (Agarwal et al., 2009) provides the right to life and liberty, as well as the freedom of expression and movement. Therefore, patients in India have the right to receive medical consultation and to be transferred to other instructions for treatment purposes. In such a scenario, the patients have to express willingness to share their data. If the hospital wants to share the patient's data for research purposes, the patient's consent is also required. If the user has also granted their consent, the data transmission operation takes a lengthy time. Furthermore, whether data is given in paper format or even by email, there are time, speed, storage, and security issues. The disadvantages of keeping data in a database include limited storage space and exposure to cyberattacks. Attackers may get access to sensitive patient information if the system is broken.

In addition, it is dangerous to use a centralized database because it is not possible to set different access restrictions for different users, search methods through encrypted communication channels, a significant amount of memory for storing and organizing medical data, and so on. Data security may be assured by employing a distributed ledger for medical data sharing in a transparent manner, similar to how Bitcoin was designed to protect transactions. Since it is medical data, patient privacy and safety must be prioritized. It is necessary to create a patient-centered access paradigm. Furthermore, patients' medical records should not be made public but should be flexible so that patients are aware of which data is shared with insurance companies and which with blood banks, etc. Researchers created a blockchain-based medical solution to avoid data theft and leaking. You may be confident that your information will be safe and secure if you use this technology. If this technique is used with cloud computing technologies, it is also feasible to eliminate five storage-related difficulties. The blockchain might also be used to fix the cloud's security flaws. By exchanging and storing medical data in a blockchain-based cloud, various issues related to medical data can be effectively addressed.

1.4 OVERVIEW OF ARCHITECTURE AND TAXONOMY OF BLOCKCHAIN

Blockchain is a distributed network-based ledger that operates on a peer-to-peer basis. Each record is included within a block sequence, which is then saved in the database. This is because each block header contains the hash of the preceding block, which is time-stamped. It enables decentralization and immutability by making the ledger accessible to all users, which makes the blockchain impenetrable to hacking. All transaction data is time-stamped, making it impossible to tamper with, and everyone has access to a copy of it. For example, when using a distributed network with a blockchain design, each member inside the network is responsible for maintaining, approving, and updating newly created records. Control of the system is exercised not just by a small number of persons but by everyone connected to the blockchain network as a whole. Each member is responsible for ensuring that all records and processes are in order, which results in the legitimacy and security of the data collected. It is possible to achieve an agreement between parties that do not necessarily trust one another.

To conclude, blockchain technology is a decentralized and distributed ledger (either public or private) of various kinds of exchanges that are structured into a peer-to-peer logical and organized network (P2P network) (Acharjamayum et al., 2018).

Even though this network contains numerous computers, the data cannot be altered without the agreement of all of the machines on the network (each separate computer).

Table 1.1 enlists some of the blockchain systems in healthcare. Blocks with transactions in a certain order are used to depict the blockchain technology's organizational structure. These lists can be kept in the form of a flat file (txt. format) or a simple database.

In general, blockchain technology employs two important data variables. (i) A data structure that holds information about the location of other variables is known as a pointer. These pointers highlight the location of certain variables. (ii) Linked lists are a series of blocks containing their data and a pointer to the next block.

Table 1.1 Summary of the few blockchain systems in healthcare

Platform name	Application	Characteristic features
Health application	Mobile technology application	Developed for insomnia and cognitive behavioral therapy
		Patients can record their health status, and concerned medical can access that medical record within minutes
Healthcare Data Gateway	Mobile technology application	Employs a simple unified indicator-centric schema (ICS) and secure multiparty computing (MPC) system for data organization
MedBlock	Electronic medical record (EMR) access portal	Improved consensus mechanism prevents the *network* from being derailed and ensures that the network is not overloaded
MIStore	Medical insurance storage system	Employ multiple servers for hospitals, patients, insurance companies, etc. to verify each other's activities to ensure security
OmniPHR	Patient health records (PHR) access and storage system	All patient medical records can be stored in one accessible forum, and different patient data sets can be incorporated into different blocks on the chain
Blockchain-based secure and privacy-preserving patient health information sharing (BSPP) scheme	Sharing of health data	Each patient's data is managed in a private blockchain system, while non-sensitive activity-related data is managed in a semi-private/consensus blockchain system
Blockchain-based multi-level privacy-preserving location sharing (BMPLS) scheme	Sharing of health data	Patient location information can be efficiently shared and requests for telemedicine medical information systems can be generated

- **Public Blockchain Architecture:** A public blockchain design doesn't need any permission implying that the data and the system are both accessible to anybody willing and able to participate.
- **Private Blockchain Architecture:** In contrast to the public blockchain architecture, the private blockchain architecture is managed solely by

participants from a particular company or individuals who have been duly requested to participate in the private blockchain architecture.

- **Consortium Blockchain Architecture:** The structure of a consortium blockchain can be composed of several different organizations. In a consortium, processes are established and regulated by the users who have been allocated to the consortium in advance
- **Hybrid Blockchain Architecture:** Hybrid blockchain architecture combines the best characteristics of both private and public blockchain, resulting in a hybrid blockchain with several advantages. Contrary to a private blockchain, where consensus is achieved by a small number of predetermined nodes, a public blockchain allows transactions to be shared around the whole network.

Blockchain metrics are usually employed to analyze the reliability, performance profile, and scalability settings of blockchain applications and to serve as a platform for the intercomparison of different versions of blockchain applications. Some of the common metrics (Zhang et al., 2017) used in blockchain technology include the following:

- Security metrics (confidentiality, authenticity, security, identity, access control, and confidentiality).
- Architectural metrics.
- Functionality metrics (number of features available, number of features available) (smart contract, interoperability).

1.4.1 Permission-less and permissioned blockchain

Permission-less and permissioned blockchain (Gemeliarana et al., 2018) are the two basic forms. The peer-to-peer network's consensus method is mostly responsible for settling disagreements. Every user can make and validate transactions, as well as contribute new blocks, in a permission-less or public blockchain. To achieve network consensus, Bitcoin employs a Proof of Work (PoW) mechanism (Puthal et al., 2018): a mathematical problem must be solved for each freshly mined block. Bitcoin's successor, Ethereum, uses both PoW and Proof of Stake (Wang et al., 2020). To participate in either approach, nodes must use resources (either computing or monetary) to produce blocks.

A permission or consortium blockchain, on the other hand, operates more like a closed ecosystem, with an extra layer of access control that precludes particular types of nodes from doing specific tasks. This signifies that the network's nodes are not all the same. As a result, they give up some decentralization to gain some degree of centralization that will allow them to enforce their aims. There are two sorts of schemes: permissioned blockchain-based techniques and permission-less approaches.

i. **Permission-Less Blockchain:** The consensus-building mechanism that blockchains employ to verify transactions and data can be accessed by anyone in a permission-less blockchain, also known as a trustless or public blockchain. Its operation is completely decentralized to a large number of unidentified parties. Permission-less blockchains have the following key features

- Complete transaction transparency.
- Open source development process.
- Absence of central authority and lack of credibility.
- Heavy use of tokens and other digital assets as incentives for involvement.

ii. **Permissioned Blockchain:** When it comes to permissioned blockchain, which is often referred to as private chains or private data centers, previously selected parties, often members of a consortium, can communicate and participate in consensus and data validation. These blockchains are somewhat decentralized in the sense that they are dispersed among known participants rather than unknown players, as is the case with permission-less networks. Digital assets and tokens are conceivable but less prevalent than permission-less (Figure 1.2).

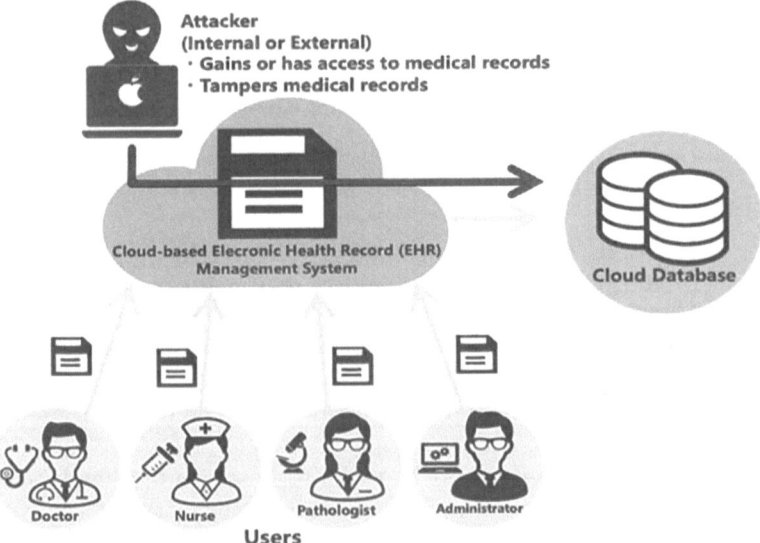

Figure 1.2 Diagrammatic sketch shows the comparison of security risks using conventional cloud-based Electronic Medical Record Management approaches.

The key features of the permissioned blockchain are as follows:
- Aims of participating organizations and controlled transparency.
- Development by private enterprises.
- The flaw in anonymity and security.
- The absence of a central authority and the authorization of decisions by private entities.

1.4.2 Cryptography for medical data sharing

Due to its cryptographic capabilities, blockchain is considered to be a new emerging technology for distributed databases that guarantees data integrity, security, and indestructibility [33]. These features make it suitable for data storage with both security and reliability. One possible solution to the aforementioned issues is to enable owner-driven security measures to protect medical data in the cloud.

As demonstrated in cloud-based data sharing schemes, traditional security measures used in the administrative domain of a single network security provider are inadequate for the exchange of medical data across the domains of multiple healthcare providers. More advanced cryptographic primitives must be developed to address these needs. The adoption of advanced and sophisticated cryptography to protect medical data exchange on cloud storage platforms is currently the subject of various intense research activities.

1.4.3 Blockchain for medical data sharing

The previous section provided an overview of the state-of-the-art technologies for protecting medical data exchange, focusing on the adoption of blockchain technology. Both permissive and open blockchains can be used to share and manage medical data, and this study highlights the potential of these technologies. Medical data exchange is fraught with concerns about security and privacy, which cannot be addressed by blockchain technology alone. We need to be aware of the limitations of blockchain technology rather than its advantages and combine it with other approaches, such as cryptographic primitives, to compensate for its shortcomings and be able to handle the security issues associated with medical information management.

Secure medical data sharing is a collaborative effort that includes patients, health care providers, and outside researchers. Healthcare providers and third-party medical researchers must comply with HIPAA's privacy and security rules (Azaria et al., 2016). To comply with HIPAA's strict privacy and security rules, medical data management must ensure the secure storage of raw medical information (confidentiality, integrity), access control, data auditability, and traceability, as well as data interoperability. In

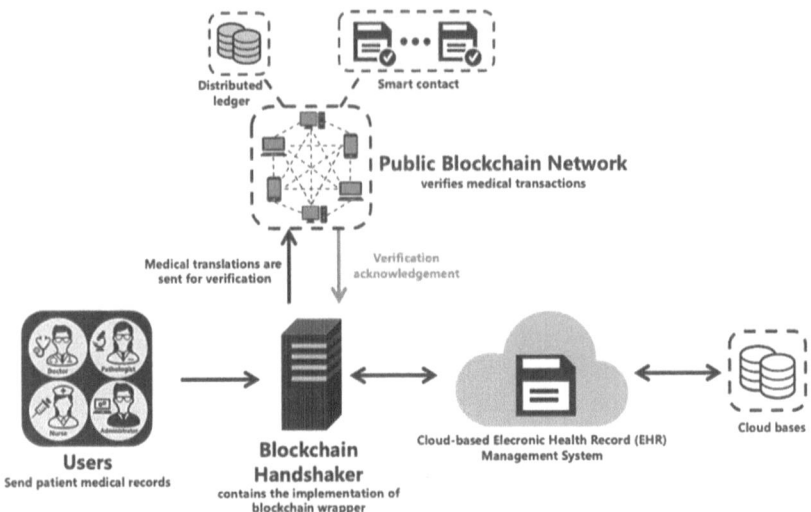

Figure 1.3 Diagrammatic sketch displays the architecture of blockchain-based electronic health record management systems.

addition, the use of blockchain for medical data exchange still requires further research on the essential elements (Figure 1.3).

Since the financial crisis of 2008, the healthcare sector has seen the most growth in terms of both annual revenue generation and data. With the proliferation of electronic health records, the need for security has never been greater. To increase the security of this vital information, the usage of blockchain technology has been urged. As a result, researchers devised a blockchain-based solution for medical healthcare that not only safeguards data from tampering but also prevents data leakage. This technology has the potential to maintain data integrity and ensure its reliability. Furthermore, if the blockchain system is properly implemented in association with cloud computing platforms, it will solve the data preservation and integrity issues. Since the cloud data is saved on multiple servers, it is widely considered one of the trusted platforms for data storage and data management. The cloud's security vulnerabilities could also be addressed using the blockchain as well. Medical data exchange and storage using a blockchain-based cloud can address several concerns.

1.4.4 On-chain or off-chain storage of medical data

Because the blockchain was initially intended to record only minute commercial transactions, it has a relatively low data storage capacity. Medical data like X-ray scans cannot be stored in Bitcoin's 1-MB block size, which is insufficient. Additionally, there are additional parts of the data cycle that must be thoroughly verified. Since the blockchain is an ever-expanding public record, on-chain data cannot be altered or erased. Because patients own their medical data, regulations such as Europe's General Data Protection Regulation (GDPR) increase patients' rights to have details of their medical records erased. The majority of data has a shelf life, which makes it unnecessary to keep them on hand indefinitely for future reference. This is also mandated by a slew of data privacy legislation.

Blockchain is a secure and transparent public digital ledger technology that can ensure the accuracy and integrity of data (transactions and blocks) recorded on the blockchain. On-chain data storage suggests that blockchain can be used to safeguard the retention of retrievable medical records. On the other hand, this naive approach requires each peer node to download the on-chain transactions and blocks locally, which results in poor performance, low throughput, and significant bandwidth waste. Most modern approaches to medical data exchange store query strings and hash values on-chain rather than off-chain, allowing for verification of authenticity and integrity. Medical data in such architecture may be protected, modified, or deleted as needed.

On-chain storing of healthcare data is not a viable option because of the restricted block size available in existing blockchains and the bandwidth consumed by the blockchain network. Off-chain storage of medical records appears to be a viable option. Off-chain storage of medical records seems to be a viable option. Blockchain technology can only secure the data held on-chain in this particular scenario, so we need to be aware of that. Therefore, in order to ensure the security and privacy of data stored off-chain, it is necessary to develop ways to store and access data using appropriate cryptographic primitives. Since each healthcare organization may be considered autonomous with its security and privacy protections, it is difficult to predict how these systems will interact.

1.5 THE BLOCK CHAIN'S LIMITATIONS

In the case of a distributed database using blockchain, extra network infrastructure and storage devices for network nodes are required to maintain the stability of the entire system. However, it aids in the conservation of human resources, the reduction of human mistakes, and the acceleration of the administrative process.

1.6 CONCLUSION

Sharing medical records while adhering to security and privacy standards has always been a challenge. Blockchain technology also speeds up the interchange of medical records and information, lowering the demand for human resources. Hospitals and clinics throughout are making it easier for people to access their medical records and other information.

This study analyses relevant solutions such as cloud-based, blockchain-based, and software-defined networking (SDN)-based techniques. For example, access and privacy controls are included under medical information security and privacy. A realistic approach to the exchange of medical data may thus require the integration of several approaches in order to meet its design objectives. Blockchain, being a novel computer paradigm, has several benefits over established technology. Permissioned or permissionless blockchains for medical data exchange are critical to picking, as we have shown here. The blockchain-based administration of medical data still has significant issues that need to be investigated and explored further. In order to shed light on these issues, we've identified several promising research avenues and approaches that might help safeguard and streamline the exchange of healthcare data.

Indeed, depending on blockchain technology to protect off-chain medical data is impossible. And, it is evident that healthcare data needed to be kept off-chain in encrypted form in order to boost network speed while still maintaining security. So, a secure healthcare system must still use cryptographic primitives to protect confidentiality, integrity, access control, and privacy. For encrypted data, complex cryptographic primitives are commonly used to impose strong and flexible encryption key access control. Encryption is expected to play a larger role in data sharing in the future.

Indeed, it is impossible to rely on blockchain technology to protect off-chain medical data. And, it is clear that in order to increase network speed while maintaining security, medical data needs to be encrypted and kept off-chain. Therefore, secure medical systems still need to use cryptographic primitives to protect the confidentiality, privacy, access control, and integrity. For encrypted data, complex cryptographic primitives are commonly used to provide strong and flexible cryptographic key access control. Encryption is expected to play a larger role in data sharing in the future.

REFERENCES

Acharjamayum, I, R Patgiri, and D Devi. Blockchain: A tale of peer to peer security. *2018 IEEE Symposium Series on Computational Intelligence (SSCI).* IEEE, 2018.

Agarwal, SS, and SS Agarwal. Medical negligence–Hospital's responsibility. *Journal of Indian Academy of Forensic Medicine* 31(2) (2009): 164–170.

Ahram, T, et al. Blockchain technology innovations. In: *IEEE Technology & Engineering Management Conference (TEMSCON)* (2017): 137–141.

Al Hamid, HA, et al. A security model for preserving the privacy of medical big data in a healthcare cloud using a fog computing facility with pairing-based cryptography. *IEEE Access* 5 (2017): 22313–22328.

Anoshiravani, A, et al. Special requirements for electronic medical records in adolescent medicine. *Journal of Adolescent Health* 51(5) (2012): 409–414.

Azaria, A, et al. Medrec: Using blockchain for medical data access and permission management. *2016 2nd International Conference on Open and Big Data (OBD)*. IEEE, 2016.

Aziz, J, et al. The impact of health information technology on the quality of medical and health care: A systematic review. *Health Information Management Journal* 38(3) (2009): 26–37.

Divya, BS and T Ramesh. Securing Data stored in clouds using privacy preserving authenticated access control. *Proceedings of IJCSMC* 3(4) (2014).

Bocek, T, et al. Blockchains everywhere – a use-case of blockchains in the pharma supply-chain. In: *IFIP/IEEE Symposium on Integrated Network and Service Management (IM)* (2017): 772–777.

Dagher, GG, et al. Ancile: Privacy-preserving framework for access control and interoperability of electronic health records using blockchain technology. *Sustainable Cities and Society* 39 (2018): 283–297.

DJabarulla, MY, et al. Blockchain-based distributed patient-centric image management system. *Applied Sciences* 11(1) (2021); 196.

Dubovitskaya, A, et al. ACTION-EHR: Patient-centric blockchain-based electronic health record data management for cancer care. *Journal of Medical Internet Research* 22(8) (2020): e13598.

Ekblaw, A, et al. A case study for blockchain in healthcare: "MedRec" prototype for electronic health records and medical research data. *Proceedings of IEEE Open & Big Data Conference* 13 (2016): 13.

Ekonomou, E, et al. An integrated cloud-based healthcare infrastructure. *2011 IEEE Third International Conference on Cloud Computing Technology and Science*. IEEE, 2011.

Fan, K, et al. Medblock: Efficient and secure medical data sharing via blockchain. *Journal of Medical Systems* 42(8) (2018): 136.

Gemeliarana, I, Gusti Ayu, K, and Sari, RF. Evaluation of proof of work (POW) blockchains security network on selfish mining. *2018 International Seminar on Research of Information Technology and Intelligent Systems (ISRITI)*. IEEE, 2018.

Griggs, NK, et al. Healthcare blockchain system using smart contracts for secure automated remote patient monitoring. *Journal of Medical Systems* 42(7) (2018): 130.

Haq, I, and Esuka, OM. Blockchain technology in pharmaceutical industry to prevent counterfeit drugs. *International Journal of Computer Applications* 975 (2018): 8887.

Iansiti, M, and Lakhani, RK. The truth about blockchain. *Harvard Business Review* 95(1) (2017): 118–127.

Ichikawa, D, et al. Tamper-resistant mobile health using blockchain technology. *JMIR mHealth uHealth* 5(7) (2017): e111.

Ida, IB, Abderrazak J, and Adlen L. A survey on security of IoT in the context of eHealth and clouds. *2016 11th International Design & Test Symposium (IDT)*. IEEE, 2016.

Jabarulla, MY, and Heung-No, L. Blockchain-based distributed patient-centric image management system. *Applied Sciences* 11(1) (2021): 196.

Jamil, F, et al. A novel medical blockchain model for drug supply chain integrity management in a smart hospital. *Electronics* 8(5) (2019): 505.

Li, H, et al. Blockchain-based data preservation system for medical data. *Journal of Medical Systems* 42(8) (2018): 1–13.

Liang, X, et al. Integrating blockchain for data sharing and collaboration in mobile healthcare applications. *2017 IEEE 28th Annual International Symposium on Personal, Indoor, and Mobile Radio Communications (PIMRC).* IEEE, 2017.

Luo, J, et al. Big data application in biomedical research and health care: A literature review. *Biomedical Informatics Insights* 8 (2016): BII-S31559.

Mani, V, et al. Hyperledger healthchain: Patient-centric IPFS-based storage of health records. *Electronics* 10(23) (2021): 3003.

McGhin, T, et al. Blockchain in healthcare applications: Research challenges and opportunities. *Journal of Network and Computer Applications* 135 (2019): 62–75.

Nakamoto, S. *Bitcoin: A Peer-to-peer Electronic Cash System* (2008).

Naresh, VS, Reddi, S, and Divakar Allavarpu, VVL. Blockchain-based patient-centric health care communication system. *International Journal of Communication Systems* 34(7) (2021): e4749.

Ølnes, S, et al. Blockchain in government: Benefits and implications of distributed ledger technology for information sharing. *Government Information Quarterly* 34(3) (2017): 355–364.

Puthal, D, et al. Everything you wanted to know about the blockchain: Its promise, components, processes, and problems. *IEEE Consumer Electronics Magazine* 7(4) (2018): 6–14.

Raghupathi, W, and V Raghupathi. Big data analytics in healthcare: Promise and potential. *Health Information Science and Systems* 2(1) (2014): 1–10.

Ray, PP, et al. Blockchain for IoT-based healthcare: Background, consensus, platforms, and use cases. *IEEE Systems Journal* 99 (2020): 1–10.

Rodrigues, JPC, et al. Analysis of the security and privacy requirements of cloud-based electronic health records systems. *Journal of Medical Internet Research* 15(8) (2013): e2494.

Roehrs, A, et al. OmniPHR: A distributed architecture model to integrate personal health records. *Journal of Biomedical Informatics* 71 (2017): 70–81.

Salah, K, et al. Blockchain for AI: review and open research challenges. *IEEE Access* 7 (2019): 10127–10149.

Shen, B, Guo, J and Yang, Y. MedChain: Efficient healthcare data sharing via blockchain. *Applied Sciences* 9(6) (2019): 1207.

Singh, R, Tanwar, S and Teek, PS. Utilization of blockchain for mitigating the distributed denial of service attacks. *Security and Privacy* 3(3) (2020): e96.

Swan, M *Blockchain: Blueprint for a New Economy.* O'Reilly Media, 2015.

Theodouli, A, et al. On the design of a blockchain-based system to facilitate healthcare data sharing. *2018 17th IEEE International Conference on Trust, Security and Privacy in Computing and Communications/12th IEEE International Conference on Big Data Science and Engineering (TrustCom/BigDataSE).* IEEE, 2018.

Usman, M, and Qamar, U. Secure electronic medical records storage and sharing using blockchain technology. *Procedia Computer Science* 174 (2020): 321–327.

Wang, HL, et al. Blockchain-based medical record management with biofeedback information. *Smart Biofeedback-Perspectives and Applications*. IntechOpen, 2020.

Wu, R, Gail-Joon, A, and Hongxin, H. Towards HIPAA-compliant healthcare systems in cloud computing. *International Journal of Computational Models and Algorithms in Medicine (IJCMAM)* 3(2) (2012): 1–22.

Zhang, P, et al. FHIRChain: Applying blockchain to securely and scalably share clinical data. *Computational and Structural Biotechnology Journal* 16 (2018): 267–278.

Zhang, P, et al. Metrics for assessing blockchain-based healthcare decentralized apps. *2017 IEEE 19th International Conference on e-Health Networking, Applications and Services (Healthcom)*. IEEE, 2017.

Zhou, L, Wang, L, and Sun, Y. MIStore: A blockchain-based medical insurance storage system. *Journal of Medical Systems* 42(8) (2018): 1–17.

Chapter 2

Industry 4.0 cross-domain blockchain solutions in healthcare

Trends and clinical considerations

Usharani Hareesh Govindarajan
University of Shanghai for Science and Technology

Gagan Narang
University of Delhi

Nikola Zivlak
Emlyon Business School

Vinay Surendra Yadav
National Institute of Technology

CONTENTS

2.1 INTRODUCTION

Industry 4.0 was conceived as a German roadmap increasing the adoption of Information Technology (IT) propelling manufacturing rate and cutting costs. The initiative has expanded to all major manufacturing economies around the world enabling smart manufacturing with added tactical intelligence. Industry 4.0 can be defined as the aggregation of a variety of technologies and techniques including Internet of Things (IoT), cyber physical systems (CPS), cloud computing, virtual reality, artificial intelligence (AI), blockchain, and big data analytics to improve the goal of a near-zero defect state. The concept strives to achieve a link between machines and systems utilized in manufacturing and develop intelligence along the value chain. Further enhancing efficiency and influencing relationships among key stakeholders. Subsequently, several regional initiatives that accommodate Industry 4.0 outline including Advanced Manufacturing Partnership 2.0 in the USA, Factories of the Future, which is the innovation vision in European Union, Made in China 2025 as the strategic plan from China, Society 5.0 connected industries in the Japan, Smart Industry in Korea, Make in India, and several others have seen considerable investments to extended manufacturing in the recent years (Shukla & Shankar, 2022). The emerging technologies in the Industry 4.0 context have gained popularity recently and are still in their infancy (Raut et al., 2020). Security and privacy are genuine worries towards application-based implementation towards Industry 4.0 (Fraga-Lamas & Fernández-Caramés, 2019). In some cases, unauthorized data breaches and information leakage may result in severe financial loss. Thus, the requirement of robust security measures is of utmost importance to deal with attacks like phishing, noise interference, Distributed Denial-of-Service (DDoS), manipulation, data rate alteration, Address Resolution Protocol (ARP) spoofing attacks, network congestion, and configuration threats to maintain integrity and confidentiality of data. Additionally, with an increased adoption trend of Industry 4.0 linked technologies, security breaches and newer cyberattacks will be observed with a higher probability (Bodkhe et al., 2020). Decentralized and cryptographic solutions plausibly through the incorporation of blockchain technologies such threats to Industry 4.0 applications can be prevented.

Blockchain technologies emerged as a frontrunner having characteristics such as smart contracts, immutable databases that are distributed across peer-to-peer P2P networks, and synchronized preventing data losses (Cole et al., 2019; Yadav et al., 2021). The term blockchain is derived from the fact that the information is stored in a block and any new information is stored by creating an additional block where all blocks are linked together forming chains called a blockchain. The working of blockchain technology is shown in Figure 2.1. Readers may refer to Alladi et al. (2019) and Bodkhe et al. (2021) for a review of the applications and use cases enabled by blockchain technologies towards Industry 4.0. Digitization is an essential element for Industry 4.0 which focuses on a networked, knowledgeable, and stable value chain. Contrary to this, Healthcare 4.0 aims to transform healthcare by focusing on the storage, aggregation, sensing, and sharing of health data (Mahajan et al., 2022). Unique use cases derived from emerging technologies such as virtual reality (Trappey et al., 2020) are already disrupting the healthcare ecosystem towards traditional treatment methodologies. The healthcare sector is undergoing rapid change using information and communication technologies. E-health provides quick access to past diagnosis through patient health records allowing effective treatments and opening new application domains of technology-based treatments under

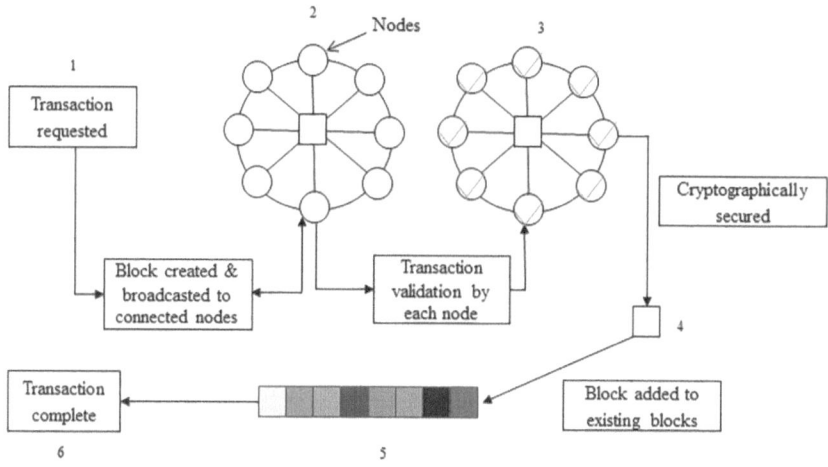

Figure 2.1 Working of blockchain. (Adapted from Yadav & Singh, 2019.)

the auspices of smart healthcare towards wellness such as extended reality (Govindarajan et al., 2021). However, ensuring a safe environment for these operations is of utmost importance since any vulnerability may be exploited by a fraudulent party. Thus, the security of health data needs significant concern. For this, the monitoring of Electronic Health Records (EHRs) supported on a blockchain system seems to be a viable solution. Blockchain is believed to be a paradigm shift unleashing a new era for decentralized data storage assisting in monitoring EHRs, effective diagnosis and securing the medical supply chain. Additionally, special emphasis is given to security in the blockchain-based system. Data on the blockchain system are kept on multiple locations and multiple devices and are secured through a complex algorithm. This approach eliminates two scenarios, i.e., loss of data at any single instance doesn't hamper the original data (Chen et al., 2019); in case of hacking at certain instances, the original data can be recovered from other servers (Zhao et al., 2018).

The integration of cyber and physical modules results in an exponential increase in research and development activities across the technology merger observed globally. A systematic keyword-based search executed across peer-reviewed journals, patents, standards, and open-source data platforms yield a total of 134,728 technical publications in the recent 6-year search period that represents the state of the art for Industry 4.0. Such volume of technology development shows relative maturity and presents stakeholders with an opportunity for cross-domain solution development considerations. Therefore, this chapter focuses on exploring the available academic publications, global patent grants, and standards to understand the current trends in blockchain-led Industry 4.0 that have a high potential to extend onto healthcare. The chapter is organized into six sections where Section 2.2

explores the background for understanding blockchain technologies and their scope in Industry 4.0. Section 2.3 presents the emerging trends through an analysis of recent global patent grants. Section 2.4 presents clinical standards followed by technical clusters in Industry 4.0 extracted from the patent analysis and discussions aligned to clinical standards as Section 2.5, and Section 2.6 presents the conclusion and cautions that need to be considered in cross-domain healthcare solution development. The chapter makes contribution in the following direction:

- The maturity of the academic documents is observed across peer-reviewed journals, patents, standards, open-source development platforms towards blockchain technologies, and smart manufacturing. The chapter extends a background for understanding the role of blockchain in the Industry 4.0 intersection.
- The chapter presents a contrast towards Healthcare 4.0-related academic and open source in comparison to blockchain and Industry 4.0.
- An understanding of the innovations and developments in the blockchain, Industry 4.0, and healthcare is comprehended analyzing the patent database. The major development trends and innovation directions are discussed.
- A preliminary review of the blockchain emerging technical standards provides the level of integration attained in current application scenario.
- A review based on patent grants classifies the cross-domain blockchain solutions in healthcare for effective clinical use cases.

2.2 BACKGROUND

The decentralized ledger forms the basis of a wide technological adaptation, infrastructures, and services. Blockchain technologies for convenience are classified to a five-layered architecture as presented in Figure 2.2. The logical classification is based on the reviews across academic journals divided as data, network, consensus, contract, and application layers. The data layer utilizes the ledger ability of blockchain where each connected node available in the network stores information. The presence of hashing algorithm ensures that the changes are mapped across all the connected blocks and any change is identified. The blocks in the data layer are generally decentralized, raising the importance of an adjoining network layer, encapsulating the network and communication-related protocols for distributed and connected systems. Distributed file systems such as InterPlanetary File System (IPFS) are utilized to address the content to uniquely identify each file in a global namespace connecting all the connected nodes (Zheng et al., 2018). The addition of new information to the chain of blocks to every node

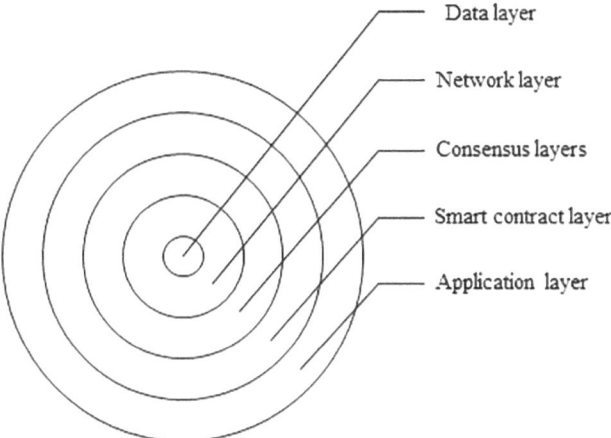

Data layer

Network layer

Consensus layers

Smart contract layer

Application layer

Figure 2.2 Logical division of blockchain architecture in layers.

is agreed upon through the infrastructure of the following layer named consensus layer. The mechanisms allow the systems to agree upon information accuracy among distributed networks. The resulting blockchain technologies allow for numerous applications across domains utilizing one or other aspects of the underlying layers from the logical classification much of which have been reviewed across several academic works of literature (Kuo et al., 2017; Yuan & Wang, 2018).

The utilization of data provenance, transparency, decentralized transaction validation, smart contracts, and immutability in the blockchain infrastructure allows for the current emerging directions for blockchain technology. A preliminary collection of document sets from global bodies including academic publishers such as the Institute of Electrical and Electronics Engineers (IEEE) Xplore, the Association for Computing Machinery (ACM), and the global citation database Web of Science (WoS). The investigation is followed by open-source development projects through popular source code management system GitHub followed by patent offices such as the China National Intellectual Property Administration, the United States Patent and Trademark Office (USPTO), and the World Intellectual Property Organization (WIPO) on a period of ranging 6 years from the year 2015 to 2021. Figure 2.3 showcases an exponential increase in technology document sets across heterogeneous platforms. There is an observable dip due to the disruptions of activities during the pandemic, while the academic documents have since then observed a rise. Blockchain distributed ledger allows for a wide area of applications and suggests a high impact on revenue, and cost which is evident in the exponential increase in technical documents over the years suspects systematic investments.

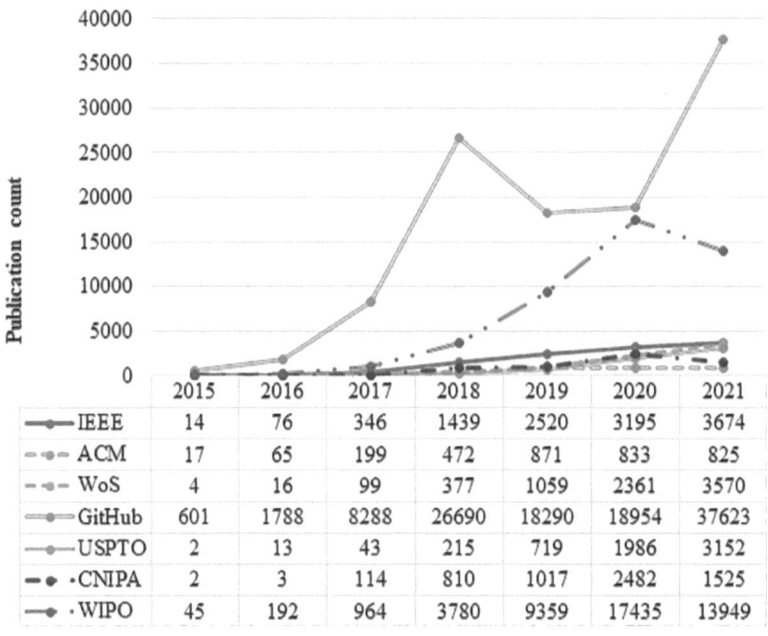

	2015	2016	2017	2018	2019	2020	2021
IEEE	14	76	346	1439	2520	3195	3674
ACM	17	65	199	472	871	833	825
WoS	4	16	99	377	1059	2361	3570
GitHub	601	1788	8288	26690	18290	18954	37623
USPTO	2	13	43	215	719	1986	3152
CNIPA	2	3	114	810	1017	2482	1525
WIPO	45	192	964	3780	9359	17435	13949

Figure 2.3 Blockchain publication trend (2015–2021).

The successful use case of Industry 4.0 technology applications is getting a lot of attention from both academia and industry. This is evident from a significant number of publications in the last decade. Furthermore, several Industry 4.0 cross-domain solutions are observed owing to the large application base of the blockchain technologies and maturity of Industry 4.0. The era of systematic change with an incorporation of smart, connected healthcare products and services is termed as Healthcare 4.0 inspired from the smart manufacturing counterpart (Tortorella et al., 2020). The systematic template used for blockchain data extraction is then used to map similar trends across Industry 4.0 and Healthcare 4.0. On a comparative analysis presented in Figure 2.4, there is a sharp contrast in the number of technical publications and open source towards Healthcare 4.0. Additionally, the patents in the Industry 4.0 segment grew without a dip during the pandemic. The Healthcare 4.0-related patents were negligible on a comparison. Bodkhe et al. (2021) surveyed to show the applications of blockchain technology for securing medical records. Similarly, Tadaka & Tawalbeh (2020) explored the application of blockchain technologies and their role in healthcare, CPS, and Industry 4.0. Jin et al. (2019) have reviewed the security and privacy measures of medical data towards digital healthcare utilization. The authors also emphasized permissionless and permissioned blockchain platforms along with their merits and demerits. The blockchain applications are related to maintaining and securing EHRs. However, few

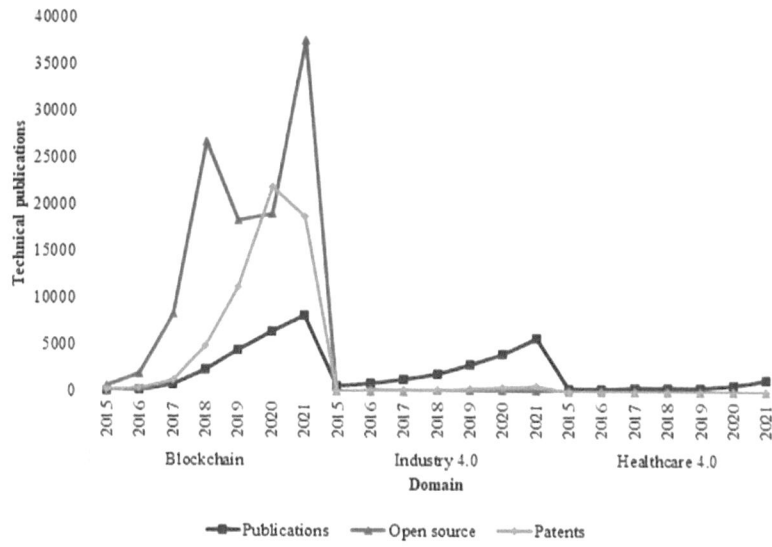

Figure 2.4 Contrast across published technical document trends cross-domains.

researchers also point out securing the critical medical supply chain. Here, in the chapter, we will limit ourselves to blockchain-based solutions for the healthcare sector. The other applications include better interoperability between healthcare databases, device tracking, maintaining hospital assets, and prescription databases, providing increased databases access (Tanwar et al., 2020). Cerchione et al. (2022) focus on the ability of blockchain-enabled technologies to revolutionize the medical field through digitizing healthcare services. The authors utilized information processing theory to design and develop blockchain-enabled EHRs for medical data storage and information sharing among the healthcare stakeholder with reduced environmental uncertainty resulting in three tangible benefits in terms of clinical, organizational, and management aspects.

However, to leverage the benefits of blockchain utility in the healthcare sector, several barriers need to be mitigated. These barriers exist at technical, organizational, legal, and governance levels. In this regard, Govindan et al. (2022) have identified 19 obstacles in the adoption of the blockchain-based platform and modeled the interrelationships between them through a methodology based on Weighted Influence Non-Linear Gauge System (WINGS). The authors claimed, "constraints resulted from financial issues", "security issues", "absence of expertise and knowledge", and "uncertain government policies" as the major barriers towards adopting blockchain platforms in the healthcare sector. Few reputed organizations like Deloitte and PwC also emphasize that regulation and legal issues are major roadblocks (Deloitte Blockchain Survey, 2018; PwC Blockchain Survey, 2018). Other studies in this respect could be referred to in the work of Gökalp et al. (2018), and Sharma & Joshi (2021). Additionally, one issue may be whether to keep medical data on-chain or off-chain. Earlier the design of blockchain was limited for small storage which is not suitable for large medical data like X-ray records, etc. Furthermore, General Data Protection Regulation (GDPR) in Europe allows patients to permanently delete their records. However, blockchain data can't be altered. This application needs second thought on-chain data storage. In addition, many times, medical data has a product life and thereafter, it becomes obsolete and is protected by law. Thus, is storing data permanently a wise solution as it unnecessarily increases the size of the network? The background presents a concise discussion of how Industry 4.0 blockchain solutions will play an active role in cross-domain healthcare applications. The broad study of the current knowledge frontier and its adjacencies is critical and is presented in the following section.

2.3 EMERGING TRENDS ACROSS GLOBAL PATENT GRANTS

The mapping of current advancements of blockchain, manufacturing, and healthcare prerequisites a data search across patent databases with a corresponding analysis to extract the industry dynamics. The section aims to gain information about practical innovations and developments using the patent database. A patent is a document that is a novel discovery and a time-bound guarantee awarded towards the commercial development of the discovery. Attributes such as industrial production, value chain networks, and market share allow evaluation of the industry. The novelty linked to the publication of the patent grant and the number of applied and approved applications is an additional important attribute of industrial dynamics (Hu, 2015). However, the assessment novelty value of the patent is done manually by domain experts, which is a critical aspect for emerging technology trends understanding. Rapid technological growth makes manual identification very challenging and other approaches exist that use automated AI-based computational approaches. The global patent used in this section data has been extracted using the patent search and analysis software IncoPat (2022) which provides wide coverage of more than 120 patenting authorities spanning across the globe. Table 2.1 tabulates the most recent grants in the last 6-year period. The section extends to the patent data analysis to extract the knowledge of critical commercial trends from the databases for a global perspective. The general trends in patent filling have observed exponential growth since 2016. China has filed the maximum filed patent applications, followed by the United States and South Korea. The general trends are understood by reviewing patents through the technical features of their content using standardized classification systems set by global agencies.

Table 2.1 Patent data: queries and the corresponding patent grants

Source	Query	Patent grants
Incopat	(TIABC=(Blockchain)) AND (PD=[20150101 TO 20211231])	64608
Incopat	(TIABC=(Blockchain) AND TIABC=(manufacturing)) AND (PD=[20150101 TO 20211231])	757
Incopat	(TIABC=(Blockchain) AND TIABC=(healthcare)) AND (PD=[20150101 TO 20211231])	327

The classification system allows the type of innovation, and the rules are standardized by the Cooperative Patent Classification (CPC) system jointly managed by the European Patent Office (EPO) and the USPTO.

CPC is an extension of the International Patent Classification (IPC) system in which each classification consists of a symbol such as "G16H10/60" attached to the patent based on the technology coverage. Using "G16H", it is evident that the patent involves technologies handling patient-related healthcare data and complete CPC code specifically for electronic patient records. The CPC system is logically divided into nine major sections using alphabets A–H and Y. The sections are subdivided into classes, subclasses, groups, and subgroups that allow for approximately 250,000 entries (European Patent Office, 2017). The classification rules ensure that each patent is assigned to at least one class; hence, there is not a one-to-one mapping of the patent grants to a code and no upper limit on the number of classes that a single patent holds. An assignee is an entity that has the current ownership rights to a patent grant. The analysis of 757 and 327 grants were observed in intersection with the total 64,608 published documents under the query blockchain. The retrieved grants were searched using the IncoPat database for patents titled or containing the keywords "manufacturing" and "healthcare" in the title, abstract, and claims section along with their intersection with the keyword "blockchain". This data is analyzed for the years 2015–2021, which is plausible since individual published documents before this period were negligible. Table 2.2 presents patent grant trends of various applicants helpful in categorizing the domain technology leaders, competition landscape, and the newcomers. Additionally, it showcases the applicant classification to understand the types of the innovation entities for technology location in the industrial chain and evaluate possible operation models. The applicant-type distribution of patents is supported only through patents from China which is the largest patent origination for around 49,785 patents in contrast with 20,414 from the USPTO in the case of blockchain. Enterprise college, personal, and research institution-based applicants by large lead the cross-domain blockchain-based manufacturing and healthcare. However, there are some

Table 2.2 Cross-domain applicant trends and types by top publishers

Domain	Assignee	Patent count	Applicant type	Patent count
Industry 4.0	General Electric Company	28	Enterprise	189
	Siemens Aktiengesellschaft	19	College	66
	Eight Plus Ventures LLC	17	Personal	29
	Univ Guangdong Technology	15	Research institute	12
	Moog Inc.	13	Government & organization	2
Healthcare 4.0	International Business Machines Corporation	17	Enterprise	23
	Nant Holdings IP LLC	13	College	8
	Janssen Pharmaceutica NV	8	Personal	2
	LBXC Co Ltd.	8	Research institute	2

patents observed from government organizations that focus on block-chain-based industrial applications. General Electric Company a multinational conglomerate incorporated in New York, United States of America dominates the individual global industry with 28 patents. Second in lead is Siemens Aktiengesellschaft, a German corporation operating globally, which has 19 patents towards Industry 4.0. Following with 17 patents is the US-based corporation Eight Plus Ventures LLC. The Guangdong University of Technology based in Guangzhou, China, holds 15 patents and is a college applicant type. Moong Inc. is a US-based aerospace and defence company specializing in precision control products and industry leveled automation towards which they have 13 patents. Focusing on blockchain cross-domain application trends in healthcare, International Business Machines (IBM) Corporation is the market leader with 17 individual patents. IBM is a major research organization with and the leading US patent applicant for almost two decades with the most number of yearly patents. The company's additional focus area is using intelligent technologies across application areas using hybrid cloud technologies, quantum computing, and cybersecurity (Bajpai, 2021). Nant Holdings IP LLC holds 13 patents followed by 8 and 9 patents by Janssen Pharmaceutica NV and LBXC Co Ltd, respectively.

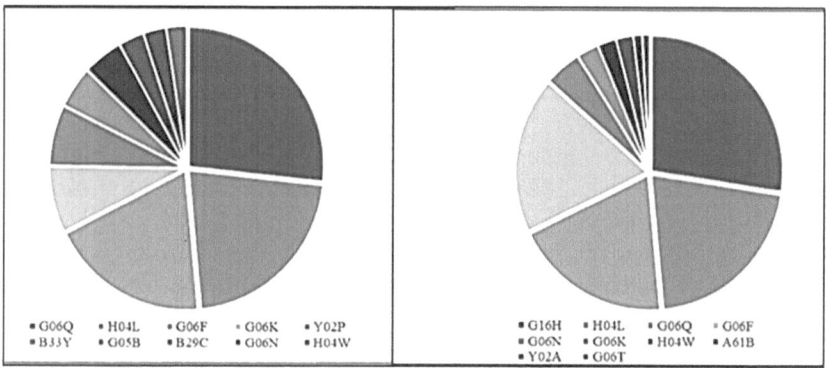

Figure 2.5 The evolution of top CPC (main) codes intersects in blockchain with Industry 4.0 (left) and healthcare (right).

Further investigations are achieved through a CPC analysis to visualize the innovation trends in the blockchain-based cross-domain applications in Industry 4.0 and healthcare.

The analysis begins the hierarchical unfolding of the patents by first examining the CPC main code to understand the broad areas of development. Figure 2.5 represents the distribution across the CPC main during the 6-year period of the collected data set. The analysis of the patent dataset in the notified period shows that maximum patents were granted to CPC code G06Q in smart manufacturing and G16H in healthcare. The code G06Q focuses on systems that process data and are designed for administrative, commercial, financial, managerial, supervisory, or forecasting purposes. This is in agreement with the general managerial consensus of research needs in data collection and processing for better administrative and managerial purposes. The code G16H focuses on healthcare informatics that includes ICT specifically adopted for the handling and processing of healthcare data. The code H04L is the second most published and is shared in both the crosssections across smart manufacturing and healthcare with blockchain and covers the methods and arrangements for digital transmission of signals and telegraphic communication. The subclass G06Q includes systems for data processing and G06N includes general computing arrangements that are based on specific computational models; the details of the model are further analyzed in the subclass. Furthermore, another code common in both explorations is the subclass G06F covers electric digital data processing and covers computing elements data structures, processing natural language, and security protocols. The subgroup G06K covers graphical data recording focusing on image or video recognition and presentation of data. The unique codes observed in the analysis for industry and blockchain systems included filings towards the code Y02P focusing on the climate change mitigation technologies and B33Y that focused on manufacturing three-dimensional objects by deposition or layering. The unique codes towards healthcare included H04W focusing on the wireless communication networks and G06T covers clinical instruments or processes for diagnostic purposes.

Table 2.3 shows the top CPC (main) codes, their ranks concerning the number of patents, and patent counts along with the corresponding CPC groups. Among the first CPC main group, the CPC group G06Q50 has the highest patent count of 188 and deals with systems that are fine-tuned for specific business purposes. These groups are found with the subgroup G06Q50/04 (manufacturing) and G06Q30/018 (technologies for e-commerce and product certification or verification). Additionally, similar CPC main code features in the healthcare and blockchain cross applications towards the subgroups G06Q2220/00 (processing utilizing cryptography in business). The CPC main code H04L2209 has the second-highest patents at 251 and deals with applications relating to mechanisms that are cryptographic and find application-securing communications. H04L9 dealing with network communication protocols for network security has 352 patents. Additionally, these main codes are also found towards healthcare with the difference in the subgroup H04L2209/88, which includes cryptographic mechanism towards medical instruments specifically. The patents with subgroups are the third highest in the analysis with G06F21 and G06F16. These deal with the security arrangements and mechanisms for hardware, software, or data against unauthorized activity, database structures, and file system structures, respectively. Other emerging areas such as blockchain-based industry 4.0 systems contributing towards greenhouse gas emission mitigation were observed through subgroup Y02P90/30 with 78 patents. The individual healthcare and blockchain patents were mainly observed focused on health informatics through majorly appearing subgroups such as G16H10 (technologies for handling healthcare data), G16H40 (technologies for the management of medical equipment), G16H15 (generation and management of medical reports), and G16H80(facilitating communication between medical practitioners). The 44 patents under the common subgroup of G04Q2220 through G06Q2220/00 focus on healthcare business processing using cryptography. The other subgroups with considerable patent inventions through G06F21/64 focused on preventing unauthorized reproduction or copying of disc-type recordable media that are still majorly used for reproduction of health records such as results of magnetic resonance imaging (MRI).

Table 2.3 Cross-domain applicant trends and types by top publishers

Cross domain	Ranking	CPC (main)	CPC (group)	Patent count (CPC group)	Patent count (CPC main)
Industry 4.0	I	G06Q: Data processing systems	G06Q50	188	251
			G06Q30	63	
	2	H04L: Transmission of digital information	H04L2209	251	603
			H04L9	352	
	3	G06F: Electric digital data processing	G06F21	87	161
			G06F16	74	
	4	Y02P: Climate change mitigation technologies	Y02P90	78	78
Healthcare	I	G16H: Healthcare informatics	G16H10	157	472
			G16H40	51	
			G16H15	44	
			G16H80	40	
	2	H04L: Transmission of digital information	H04L2209	115	249
			H04L9	134	
	3	G04Q: Data processing systems	G04Q2220	44	44
	4	G06F: Electric digital data processing	G06F21	89	89

2.4 EMERGING TECHNICAL STANDARDS

The adoption of the latest technologies with healthcare requires understanding both the healthcare sector and the impact of technology ensuring fulfillment of the required medical needs. The process of integrating technology raises concerns and insecurities such as data privacy, regulation, proper technology implementation, monitoring of the technology's negative effects, and so on. Addressing such concerns, standardization aids globally interoperable technologies, thereby making the underlying technology more accessible and affordable. This additionally establishes trust in the technology in the users and manufacturers globally. The Institute of

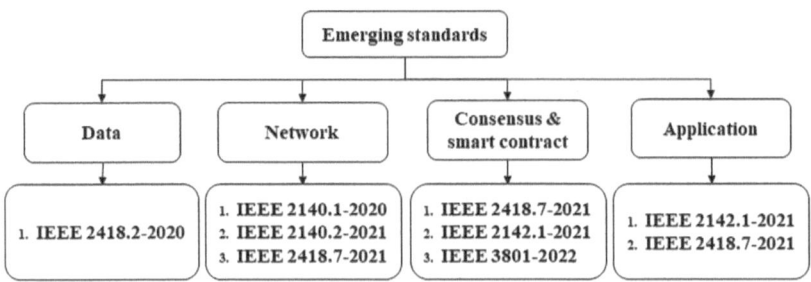

Figure 2.6 The emerging standards (active) mapped towards blockchain layers.

Electrical and Electronics Engineers Standards Association (IEEE SA) is an association that develops global standards for many industries, including healthcare (IEEE Standards Association, 2021). The standards are adopted by stakeholders leading to a global consensus mechanism on the offered technologies, products, and services. Figure 2.6 presents the recent and emerging standards curated towards the logical layer division. The compilation is done through a review in the IEEE SA database. Standards are logically split into the layer structure from the previous section. Since the standards are still emerging, the consensus and smart contract layer are merged and the compilation is one to many mappings presenting an architectural framework defining functional roles for blockchain-driven systems.

The standards towards the data layer establish the data format requirements for blockchain systems. Additionally, they serve as the reference for the data structures, data types, and data elements stored systems. The prominent standard in the category is the IEEE 2418.2-2020. The blockchain systems are decentralized and the network layer includes covers aspects of establishing connections, security management towards data exchanges, security, authentication, protection of the user from unanimous access. The standard IEEE 2140.1-2020 and its recent version IEEE 2140.2-2021 along with the IEEE 2418.7-2021 address such issues. The consensus and smart contract encapsulate the standard rules towards the addition of new information to the blockchain system and automated execution towards a certain set of rules. IEEE 2418.7-2021 defines functional roles for blockchain-driven supply chain finance (SCF) implementations that focus on their applications in financing institutions. These are used to attain functionalities such as asset issuance and transfer, loans based on the issued asset stored on chain. Further includes the asset management aspects such as clearing and settlement, and tracking. Further, the IEEE 2142.1-2021 features standardize the e-invoicing generated in businesses that are based on the blockchain system. It features typical business functionalities through digital platforms and related security mechanisms. The application layer includes the implementation of blockchain technologies towards fully functional systems. The standards in the layer are not unique to the layer and feature in other layers with a contrast being in the layer the standardization is on the techniques for the requirement of the business system, and blockchain platforms. The list of standards added to the layer includes IEEE 2418.7-2021 and IEEE 2418.7-2021.

2.5 TECHNOLOGY CLUSTERS AND CLINICAL ASPECTS

This section logically extends systematic analytics of the collected patent grants data from the IncoPat global database investigating the research direction towards utilization blockchain and smart manufacturing-based solutions in the healthcare industry. VOSviewer is utilized to extract the key terms from the corpus of academic documents and extract information, as shown in Figure 2.7. VOSviewer is a software tool utilized for the visualization of patent grants (Van Eck & Waltman, 2013). Text mining is attained through the tool which creates co-occurrence maps based on textual data from the abstract and titles of the collected corpus. The objective has been to extract keyword patterns and assess the connections between terms by creating a network visualization map. The map is automatically created where the back-end implementation involves two major stages term identification and term selection. For the term identification stage, the software uses the Apache OpenNLP library to preprocess text data. In the term selection stage, users can customize the word list. The investigation included the scan of around 917 terms that occur at least five times, and for each of the 917 terms, a relevance score was calculated. The most relevant terms are selected for the analysis of the patent applications. The connection between the nodes indicates a disjoint from the smart manufacturing sector sporadically connected with healthcare-related terms.

The selected four key observed clusters concerning the healthcare sector selected based on the keywords include digital healthcare, implants man and machine interface, precision surgical tools, and clinical care and informatics, which are analyzed further for deep dive.

Digital healthcare and informatics cluster is identified as a key destination where most of the patents in the current time are filed. Digitalizing healthcare is very critical, especially in the current time frames. Focusing primarily on understanding healthcare users' requirements and fulfilling them utilizing the underlying digital infrastructure. Informatics platforms for collecting and sharing of healthcare data, use of mobile apps and wearables for patient surveillance for prevention of health issues, and monitoring patients in the preoperative period, providing education related to health and self-service of the health evaluation system and intelligent claims process. Major patents granted in this cluster are towards blockchain-based systems that

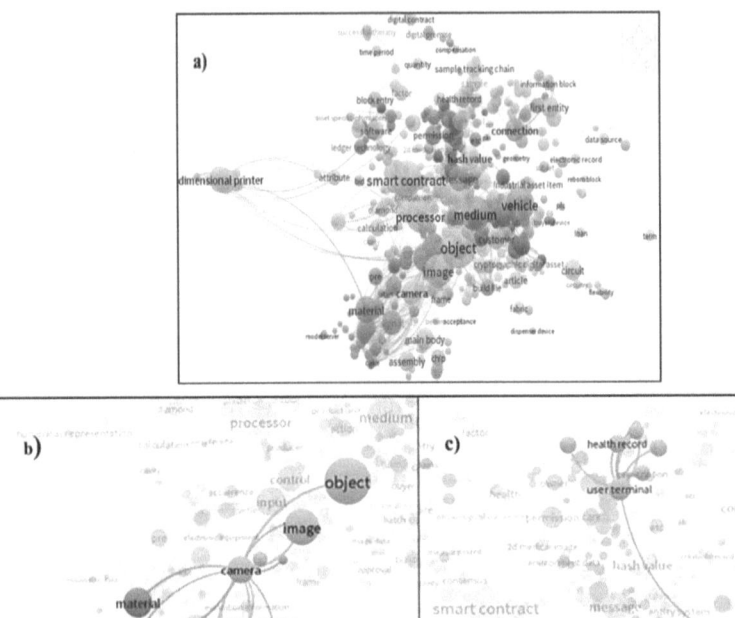

Figure 2.7 (a) The network visualization plot of patent grants data. (b) Keyword cluster connected across smart manufacturing related to the blockchain. (c) Keyword cluster associated with user-based applications in healthcare.

propel the digitization of healthcare infrastructure through innovation. An example case is US Patent grant US20190266597A1 granted to Panaxea Life Inc., which discloses a computer-implemented method for executing healthcare transactions. The novelty of the method is in aiding the registration of a healthcare user and a healthcare provider with a healthcare service application. The registered stakeholders are communicatively coupled to a blockchain, through a respective healthcare service account. Further, the application ensures the option of depositing healthcare cryptocurrency to the healthcare service account of the user. The patent US20200402629A1 granted to the Electronic Health Record Data Inc includes a blockchain system based on Electronic Health Records (EHR) entries. The system is efficiently configured storing multiple entities within the healthcare provider ecosystem (e.g., pharmacy industry entities, traders, service providers, and healthcare users). Further, the system connects such entities enabling a convenient one-stop solution for the following communication ensuring uniformity. The patent further proposed data fetching application programming interface (API) and financial transaction API between the entities in the ecosphere. This ensures systematic centralization of the various aspects of the healthcare process through digitally connected and unified data schemas.

Implants cluster focuses on using blockchain and manufacturing-based techniques such as QR codes or emerging visual technologies for improving healthcare infrastructure. The published patent grant US20210225493A1 brings accountability to the industry for surgical implants. The patent provides transparency and safety when handling surgical devices and ultimately reduces costs for members, surgeons, and insurers. The patent grant IN202111000130A was filed at the Indian Patent Office by Eswasthalya Healthcare Private Limited including a customizable virtual hospital platform system for virtual delivery of healthcare services and education akin to physical hospital. The components leverage emerging technologies common in smart manufacturing setups to help effective and continued doctor surgical-based education.

Clinical care comes across as an important field of patent grants primarily encapsulated patents towards personalized care systems for personal or clinical applications. The use of mobile apps and wearables for patient surveillance for prevention of health issues and monitoring patients in the preoperative period, providing education related to health and self-service of the health evaluation systems. IP Australia has granted AU2020101946A4 patent to secure flexible access to the healthcare information resources (HIR) in the EHR database. The invention creates an EHR analysis presenting a summary report. The produced report adds an opportunity for including treatment strategies and ideation for the physician or healthcare user. Further, the patent AU2020102115A4 improves clinical trials and healthcare research and development. The method and technology receive the old and new profile information for a patient at a database and local server connected to a computer network from a fixed IP address and identification number; the patient profile information is then fetched by using blockchain technologies and submitted by a user at a terminal connected to the network through a comparison based on the criteria for clinical trials.

Enhancers cluster includes lifestyle and health-enhancing inventions towards personalized, societal, and clinical supported methods. For instance, the patent IN202021030276A invented a combination with blockchain and IoT-enabled monitoring systems that store the sonography of expected deliveries to prevent female feticide. The patient information is stored in the electronic patient database where sonography records are encrypted and decrypted using blockchain to store internally or remotely using blockchain-based networks. The remote storage can be accessed by law enforcers for analysis and monitoring. The patent US20190358428A1 by the applicant Xiamen Brana Design Co Ltd presented a blockchain and IoT-based robot used towards sleep environment monitoring, physiology monitoring, sound, lighting and electricity control, a smart storage compartment, central data processing, and machine arms. The IoT system senses and executes instructions from the sleep-caring robot, thereby catering to the bedroom activities of the user. The grant AU2021100088A4 used a Multimedia Data Digest Crypto Blockchain-based Chaotic Ordered (MDDC-BCO) method and effectively secured communication between the connected healthcare devices. The patent grant IN201742024550A strives to securely provide an ecosystem for Unique Digital Medical Record registration services and birth registration services.

2.6 CAUTIONS AND CONCLUSIONS

The Covid-19 pandemic exposed the weaknesses in the globally interconnected supply chains manufacturing-related direct impact on the healthcare infrastructures. The chapter reviews the exponential growth observed in the blockchain technologies and Industry 4.0 manufacturing initiative emphasizing technological innovation through the patent data analysis additionally studying the role of IT helping the manufacturing companies upgrade infrastructure. Industry 4.0 linked high-precision devices and blockchain-enabled information and communication technologies improve business processes and competitiveness, therefore are at a stage to develop cross-domain solutions across the healthcare sector. Such technologies provide the ability to deliver the required resources and integrate the data needed for rapid decisions making. The healthcare sector is undergoing rapid change pushing the management towards discovering new clinical use cases for emerging technologies. The chapter established the explosive growth in blockchain, and Industry 4.0 linked technological published works in contrast to healthcare. Industry 4.0 linked smart manufacturing strongly encourages computer-supported systems by which the medical field can take significant advantage upon their adoption. The high-precision techniques allow for high-quality medical devices and securing the infrastructures using blockchain-based solutions to meet the demand of healthcare users and reach the goals of Healthcare 4.0. The upgradation of medical equipment through advanced manufacturing equipment and securing healthcare records using immutable blockchain-based medical fields were found to be the major emerging cross-domain applications areas. Additionally, such cross-domain applications have a better impact on the environment by reducing clinical biowaste and ultimately making the healthcare ecosystem more agile improving the business processes. Industrial and medical stakeholders can utilize the key identified areas towards the incorporation of blockchain technologies across healthcare project development and workflow. The major concern and future direction for the work include the study of adoption barriers and the establishment of business models that provide the already strained healthcare sector with security on returning investments. The investigation additionally recommends industrial and healthcare managers primarily focus on the specific combination of wearable devices with underlying blockchain standards and expand towards a strategic asset in the future.

REFERENCES

Alladi, T., Chamola, V., Parizi, R. M., & Choo, K. K. R. (2019). Blockchain applications for industry 4.0 and industrial IoT: A review. *IEEE Access*, 7, 176935–176951.

Bajpai, C. P. (2021). *Top Patent Holders of 2020. National Association of Securities Dealers Automated Quotations.* https://www.nasdaq.com/articles/top-patent-holders-of-2020-2021-01-29.

Bodkhe, U., Tanwar, S., Bhattacharya, P., & Verma, A. (2021). Blockchain adoption for trusted medical records in healthcare 4.0 applications: A survey. In *Proceedings of Second International Conference on Computing, Communications, and Cyber-Security* (pp. 759–774). Springer, Singapore.

Bodkhe, U., Tanwar, S., Parekh, K., Khanpara, P., Tyagi, S., Kumar, N., & Alazab, M. (2020). Blockchain for industry 4.0: A comprehensive review. *IEEE Access*, 8, 79764–79800.

Cerchione, R., Centobelli, P., Riccio, E., Abbate, S., & Oropallo, E. (2022). Blockchain's coming to hospital to digitalize healthcare services: Designing a distributed electronic health record ecosystem. *Technovation*, 102480.

Chen, L., Lee, W. K., Chang, C. C., Choo, K. K. R., & Zhang, N. (2019). Blockchain based searchable encryption for electronic health record sharing. *Future Generation Computer Systems*, 95, 420–429.

Cole, R., Stevenson, M., & Aitken, J. (2019). Blockchain technology: implications for operations and supply chain management. *Supply Chain Management: An International Journal*, 24(4), 469–483.

Deloitte Blockchain Survey, 2018. Available on: https://www2.deloitte.com/content/dam/Deloitte/cz/Documents/financial-services/cz-2018-deloitte-global-blockchain-survey.pdf.

European Patent Office. (2017). *EPO – Cooperative Patent Classification (CPC)*. 2007 European Patent Office. All Rights Reserved. https://www.epo.org/searching-for-patents/helpful-resources/first-time-here/classification/cpc.html.

Fraga-Lamas, P., & Fernández-Caramés, T. M. (2019). A review on blockchain technologies for an advanced and cyber-resilient automotive industry. *IEEE Access*, 7, 17578–17598.

Gökalp, E., Gökalp, M. O., Çoban, S., & Eren, P. E. (2018). Analysing opportunities and challenges of integrated blockchain technologies in healthcare. In *Eurosymposium on Systems Analysis and Design* (pp. 174–183). Springer, Cham.

Govindan, K., Nasr, A. K., Saeed Heidary, M., Nosrati-Abargooee, S., & Mina, H. (2022). Prioritizing adoption barriers of platforms based on blockchain technology from balanced scorecard perspectives in healthcare industry: A structural approach. *International Journal of Production Research*, 1–15. https://doi.org/10.1080/00207543.2021.2013560.

Govindarajan, U. H., Zhang, D., & Anshita. (2021). *Extended Reality for Patient Recovery and Wellness. Extended Reality for Healthcare Systems: Recent Advances in Contemporary Research.* [Print ISBN: 9780323983815, Electronic ISBN: 9780323985390]

Hu, M. C. (2015). Industrial dynamics: Patenting perspectives. *International Encyclopedia of the Social & Behavioral Sciences*, 834–842. https://doi.org/10.1016/b978-0-08-097086-8.73095-9.

IEEE Standards Association. (n.d.). *IEEE SA – What Are Standards?* Retrieved July 25, 2021, from https://standards.ieee.org/develop/develop-standards/overview.html.

Jin, H., Luo, Y., Li, P., & Mathew, J. (2019). A review of secure and privacy-preserving medical data sharing. *IEEE Access*, 7, 61656–61669.

Kuo, T. T., Kim, H. E., & Ohno-Machado, L. (2017). Blockchain distributed ledger technologies for biomedical and health care applications. *Journal of the American Medical Informatics Association*, 24(6), 1211–1220.

Mahajan, H. B., Rashid, A. S., Junnarkar, A. A., Uke, N., Deshpande, S. D., Futane, P. R., ... Alhayani, B. (2022). Integration of Healthcare 4.0 and blockchain into secure cloud-based electronic health records systems. *Applied Nanoscience*, 1–14. https://doi.org/10.1007/s13204-021-02164-0.

PwC Blockchain Survey, 2018. Available on: https://www.pwccn.com/en/research-and-insights/publications/global-blockchain-survey-2018/global-blockchain-survey-2018-survey-highlights.pdf.

Raut, R. D., Gotmare, A., Narkhede, B. E., Govindarajan, U. H., & Bokade, S. U. (2020). Enabling technologies for Industry 4.0 manufacturing and supply chain: concepts, current status, and adoption challenges. *IEEE Engineering Management Review*, 48(2), 83–102.

Sharma, M., & Joshi, S. (2021). Barriers to blockchain adoption in health-care industry: An Indian perspective. *Journal of Global Operations and Strategic Sourcing*, 14(1), 134–169.

Shukla, M., & Shankar, R. (2022). An extended technology-organization-environment framework to investigate smart manufacturing system implementation in small and medium enterprises. *Computers & Industrial Engineering*, 163, 107865.

Tadaka, S. M., & Tawalbeh, L. A. (2020). Applications of blockchain in healthcare, industry 4, and cyber-physical systems. *2020 7th International Conference on Internet of Things: Systems, Management and Security (IOTSMS)*, 1–8.

Tanwar, S., Parekh, K., & Evans, R. (2020). Blockchain-based electronic healthcare record system for healthcare 4.0 applications. *Journal of Information Security and Applications*, 50, 102407.

Tortorella, G. L., Fogliatto, F. S., Mac Cawley Vergara, A., Vassolo, R., & Sawhney, R. (2020). Healthcare 4.0: trends, challenges and research directions. *Production Planning & Control*, 31(15), 1245–1260.

Trappey, A. J., Trappey, C. V., Chang, C. M., Shih, X. Y., Govindarajan, U. H., Gupta, N., & Su, I. A. (2020). Behavioral therapy for phobias using immersive virtual reality technology. *Journal of the Chinese Society of Mechanical Engineers*, 41(2), 131–140.

Van Eck, N. J., & Waltman, L. (2013). VOSviewer manual. *Leiden: Univeristeit Leiden*, 1(1), 1–53.

WIPO Inspire. (n.d.). *IncoPat Global Patent Database*. Retrieved February 2, 2022, from https://inspire.wipo.int/incopat-global-patent-database.

Yadav, V. S., & Singh, A. R. (2019). A systematic literature review of blockchain technology in agriculture. In *Proceedings of the International Conference on Industrial Engineering and Operations Management* (pp. 973–981).

Yadav, V. S., Singh, A. R., Raut, R. D., & Cheikhrouhou, N. (2021). Blockchain drivers to achieve sustainable food security in the Indian context. *Annals of Operations Research*, 1–39. https://doi.org/10.1007/s10479-021-04308-5.

Yuan, Y., & Wang, F. Y. (2018). Blockchain and cryptocurrencies: Model, techniques, and applications. *IEEE Transactions on Systems, Man, and Cybernetics: Systems*, 48(9), 1421–1428.

Zhao, H., Bai, P., Peng, Y., & Xu, R. (2018). Efficient key management scheme for health blockchain. *CAAI Transactions on Intelligence Technology*, 3(2), 114–118.

Zheng, Q., Li, Y., Chen, P., & Dong, X. (2018). An innovative IPFS-based storage model for blockchain. In *2018 IEEE/WIC/ACM International Conference on Web Intelligence (WI)* (pp. 704–708). IEEE.

Chapter 3

Blockchain in banking and financial services

Himanshu Falwadiya, Sanjay Dhingra, and Shelly Gupta

Guru Gobind Singh Indraprastha University

CONTENTS

DOI: 10.1201/9781003282914-3

3.1 INTRODUCTION

Banks and financial institutions are the powerful pillars for the progress and development of any economy in the world. They play a crucial role in achieving the objectives of financial inclusion by providing financial services in a cost-effective manner. According to World Bank (2018), financial inclusion means that "individuals and businesses have access to useful and affordable financial products and services that meet their needs – transactions, payments, savings, credit and insurance – delivered in a responsible and sustainable way". However, banks and financial institutions face various challenges to provide equitable and quality financial services to their customers, including operational inefficiencies, bad customer experience, high processing costs, increasing fraudulent attacks, and transparency issues (Deloitte, 2017; Guo & Liang, 2016). All these challenges are reflected in the daily trade finance process as delayed payments, extended deadlines, multiple versions of the truth, manual contract creation, and duplicative bills of lading (Deloitte, 2015). As a result, banks and financial institutions are keen to adopt new technological advancements to overcome these challenges and pain points, including blockchain, mobile banking, cyber security, cloud computing, big data, and artificial intelligence. Blockchain technology has been heralded as a leading technology that can transform the entire finance platform with increased efficiency and security. The technology has offered to provide innovative, highly secure, and faster transactions at a low cost to ease financial activities performed at the domestic and international levels (Lee & Shin, 2018).

Blockchain technology is a distributed ledger that can securely maintain and verify records and reduce the double-spending in transactions (Daluwathumullagamage & Sims, 2021). According to Du et al. (2019), "A blockchain is a chain of data blocks created to record a transaction". The transactions added in the different blocks are validated by the nodes in the network using the consensus mechanism (Ali et al., 2020). The consensus mechanism, along with the decentralised and distributed nature of the ledger, helps remove the intermediaries for validating and authenticating transactions (Patki & Sople, 2020).

Though blockchain technology is being well received, it is still at an early stage. According to Jagtap et al. (2019), in 2018, blockchain technology had a value of $277.1 million in the Banking, Financial Services and Insurance (BFSI) sector and is expected to reach $22.46 billion by 2026. India and other countries are moving towards digitalisation, so this is the right time to understand the various use cases of blockchain technology and fully exploit the benefits of this innovation.

The adoption of blockchain in the banking industry will recover the trust and confidence among the stakeholders and rejuvenate the country's economic growth. To exploit the potential of blockchain technology in banking and financial services, it is essential to explore and understand its advantages, risks, and applications. Therefore, this study aims to identify the advantages, risks, and applications of blockchain technology in banking and financial services and provides an overview of the global scenario of blockchain adoption in banking and financial services.

The study has contributed to the current body of knowledge by extending the research on the blockchain through

a. Exploring critical areas explaining the benefits and challenges of adopting blockchain technology in the context of the banking and financial sector.
b. Identifying various applications and use cases where blockchain can solve multiple inherent challenges of traditional banking and financial services.
c. Outlining several blockchain-based consortiums that can serve as benchmarks in future collaborations.

The rest of the chapter is structured as follows: Section 3.2 highlights the various benefits of blockchain technology, followed by its multiple applications in the banking and financial sector in Section 3.3. Section 3.4 outlines the overall challenges present in implementing the technology. Next, Section 3.5 provides an overview of the global scenario of blockchain adoption in banking and financial services, and finally, Section 3.6 extends the concluding remarks.

3.2 BENEFITS OF BLOCKCHAIN TECHNOLOGY

Blockchain technology is highly acknowledged for its benefits as a secure, transparent, and immutable innovation within the banking and financial industry (Gan et al., 2021). The technology holds the potential to revolutionise the financial world by streamlining and automating everything from retail banking to asset trading, security issuance, and clearing and settlements. Blockchain technology provides answers to many inherent challenges of the banking and financial sector, some of which are listed below.

3.2.1 Lowering costs

One of the leading benefits associated with blockchain technology is the reduction in transaction costs (Gan et al., 2021) as it facilitates fast and secure transactions without the interference of a third party. All the parties involved in dealing can assess the information through distributed ledger approach, which eliminates the need for an intermediary to coordinate the data and legitimise the transactions. This leads to a reduction in the cost of third-party fees as well as the cost of data management. Also, blockchain technology helps individuals to obtain great exchange rates in cross-border remittances due to the real-time settlement of transactions (Gupta & Gupta, 2018). As reported by Daluwathumullagamage and Sims (2021), blockchain technology helps to reduce the cost of remittances to 2%–3%, which is approximately 5%–20% when held by third-party intermediaries.

3.2.2 Data privacy and security

As all businesses are gradually shifting to digital mode, the online platform is becoming a breeding landscape for hackers and scammers. In such a scenario, blockchain can be a risk management tool by monitoring the accessibility of information. The exchange of information among the parties involves using cryptographic algorithms that confide in public and private decryption codes to protect against fraud and unauthorised movements. The decentralised nature of blockchain assures that hackers do not have a central node of failure and are, therefore, able to defend against attacks more effectively without threatening the entire system (Gan et al., 2021). Since the data is distributed across multiple nodes, hackers do not have any single point to modify, destroy, or attack any transaction information. Also, the instant settlement of transactions reduces the risk of capturing transaction data or diverting payment as data is inalterable once verified.

3.2.3 Transparency

The term transparency is automatically related to blockchain technology because the transaction data is uniformly recorded at multiple locations, and the same information can be traced by any party of the network with permissioned access with a real-time view. The fact that all the information is immutably documented and is time-stamped adds to the transparency and guarantees a single version of the truth. The data available on the blockchain is tamper-proof, and the automation of accounting and reporting warranted by smart contracts helps trace business activities (Gan et al., 2021). This way, financial institutions can easily track the complete history of a transaction and, therefore, expose inefficiencies and virtually eliminate any possibility of a scam, money laundering, and other fake operations.

3.2.4 Disintermediation

The idea of blockchain technology is based on decentralisation that enables banks and other financial institutions to perform transactions without the intervention of a trusted third party or a central authority. Blockchain creates a shared platform by allocating control to validate the transactions to all the authorised parties in the transaction chain instead of everything being regulated by a single command. The elimination of the third party not only fastens the transaction but also mitigates counterparty risks. The individuals are assured of the transactions as the technology provides a reliable alternative for validating data. Moreover, direct communication between the parties reduces operational costs and increases trust.

3.2.5 Operational efficiency

Currently, many financial activities and banking operations are relatively slow and take a long time for approval, management, and processing. Blockchain can efficiently resolve these concerns by performing instant verification and authorisation and avoiding the possibility of duplication and errors. Also, the technology shortens the level of intermediaries to streamline financial operations. The elimination of intermediaries reduces the time of transactions to minutes and that of settlements to seconds. Since blockchain allows financial institutions to save the data in blocks, it advances the mobility of information and avoids any delays caused by documentation and duplication. In short, blockchain technology improves the operational efficiency of these organisations and enhances consumers' experiences.

3.2.6 Reduce error solving and reconciliation

Immutability is a distinguishing element of blockchain technology, i.e., once the information is entered in a block, it can't be changed retroactively. Also, all the recorded information leaves behind a complete audit trail that can be monitored in real time. This allows banks and other financial institutions to quickly reconcile the transactions, coordinate with other shareholders, and locate errors at the earliest. This way, they have accurate time and means to mend the mistakes that can create a problem for the organisation and its customers.

3.3 APPLICATION OF BLOCKCHAIN TECHNOLOGY IN BANKING AND FINANCIAL SERVICES

Blockchain technology is disrupting the business models of various industries, including banking & finance, supply chain, manufacturing, healthcare, and energy. The most famous domain of the blockchain is the banking and financial sector, as it can solve various challenges faced by banking and financial institutions. Some of the most widely used applications of blockchain technology in banking and financial services are provided in the section below.

3.3.1 Know your customer (KYC)

KYC is an essential guideline that each bank and financial institution must follow. Banks and financial institutions must follow specific customer identification procedures to open and monitor accounts. The main objective of KYC provision is to prevent the banks from fraud and criminal activities such as terrorist financing activities and money laundering (RBI, 2010). Currently, banks collect the data from the customers and upload it to their central database, which generates various challenges. The cost for performing this procedure is relatively higher, and the data is not updated on a real-time basis. According to Institute for Development and Research in Banking Technology (IDRBT, 2017), the banks are concerned due to an increase in the cost of regulatory compliance such as anti-money laundering and KYC. Blockchain technology can solve these challenges by streamlining and automating the KYC process. Banks and other authorised organisations can use the same data without bothering the customers by repeating the same procedures. The blockchain-based KYC process would help eliminate the manual processing of data, thereby reducing the administrative costs and duplicating records.

3.3.2 Trade finance

Trade finance includes various financial activities that the buyers and sellers perform related to international trade and commerce. Even today, the process of trade finance is complex and requires heavy paperwork for invoices, bills of lading, and letters of credit. Blockchain technology could simplify these procedures if all the participants of the transaction get on board and share information on a private distributed ledger. This way, trade deals can be executed quickly via several smart contracts upon fulfilling the prescribed conditions. Blockchain technology would significantly improve efficiency and transparency, error-free documentation, fast process, and reduced trade finance costs. By implementing blockchain in their operations, "Barclays and an Israel-based start-up company" have completed a business dealing in less than 4 hours which usually takes 7–10 days (Gupta & Gupta, 2018).

3.3.3 Accounting and auditing

Blockchain is a promising technology for accounting and auditing. Various accounting firms, including the big four, have expressed their interest to understand the potential of blockchain technology and its use cases in finance and accounting. Currently, firms use a double-entry bookkeeping system to record transactions. But there is a need to improve the existing reporting system. So triple entry accounting comes into the picture, recording three entries for each transaction (Grigg, 2005). These entries were initially stored by a trustworthy intermediary, which can now be performed by blockchain technology. Through this approach, all the accounting entries will be cryptographically secured by a third entry, thus leaving no space for fraud and manipulation. Once a transaction is entered into the network, the other party to the transaction can look at a particular entry and have it automatically recorded in its book (Rana, 2020). As blockchain is highly encrypted and decentralised, it saves the time of bookkeepers and ensures the authenticity of information as the data cannot be altered, edited, or removed once entered. According to Yu et al. (2018), in the short run, blockchain technology could solve the trust issues with the investors by using blockchain as a platform for disclosing financial information, whereas in the long run, it could help mitigate the errors in the firm's disclosure and improve the quality of accounting information.

3.3.4 Digital currency

Cryptocurrency and Central bank digital currency (CBDC) are the two virtual currencies that use the blockchain as an underlying technology.

The first cryptocurrency, called Bitcoin, was developed in 2008 when Satoshi Nakamoto mentioned this in his paper. Bitcoin is a digital currency that can be used as a medium of exchange and to make peer-to-peer transactions (Nakamoto, 2008). Presently, cryptocurrency is primarily utilised for investing and trading purposes. More than 8,000 cryptocurrencies are traded in exchanges globally (CoinMarketCap, 2021). The most widely well-known cryptocurrencies are Bitcoin, Ethereum, Binance coin, Tether, and Solana.

CBDC can be defined as a digital currency which can be converted or exchanged at par with similarly denominated cash and traditional central bank deposits of a nation. CBDC is a digital token of a country's official currency issued by a central bank. There are two types of CBDC: wholesale and retail CBDC.

3.3.5 Cross-border payments

Cross-border payments refer to the financial transactions between two parties from different jurisdictions. Traditionally, transferring money across borders is very time-consuming, expensive, and less transparent. Also, the clearing process for cross-border payments is different for every country and requires 3 days to process, which results in low efficiency (Guo & Liang, 2016). Blockchain technology could significantly reduce and overcome all the current pain points and improve the efficiency of the system. When used for cross-border payments, it will process transactions on a real-time basis with less cost and a 24×7 hour facility (Qiu et al., 2019).

3.3.6 Taxing

The taxing system for any country should be efficient and effective. If the taxing system is flawed, then the citizens of that particular country do not comply with its rules and regulations. Various countries face various pain points, including transparency and corruption in the current taxing system (Persson et al., 2017). The study done by Kabir (2021) found that transparency and effectiveness are low in the Bangladesh taxing system, leading to low tax collection in the country. These issues erode the trust among the citizens towards the government services. Governments from different countries are focusing on adopting blockchain technology to solve these challenges. Since the data recorded in blockchain is immutable and temper free, it will improve the transparency and better management of the taxing system.

3.3.7 Capital markets

The capital market provides a platform where the companies raise funds through various instruments (securities), and after listing on stock exchanges, the securities are traded in the secondary market. The traditional stock exchange process includes various intermediaries and involves several stages, resulting in lengthy settlement cycles, high transaction costs, and vulnerability to attacks. The adoption of blockchain technology in the capital market could provide various benefits, including fast clearing & settlement, a decline in systematic risk (Buehler et al., 2015), low cost for transactions, and enhancement of automation and transparency (Noble & Patil, 2021). Blockchain technology can eliminate unnecessary intermediaries and facilitate simultaneous transactions on exchanges worldwide. Additionally, it could be used to issue securities, significantly reducing marketing and disclosure costs. Blockchain can solve many inherent capital market challenges with its use cases, such as digital identity, corporate action, reporting, and peer-to-peer payments and transfers (Sandeep, 2021).

3.3.8 Insurance

Blockchain technology has the potential to revolutionise the insurance industry in various aspects. The insurance products offered by insurance companies are very complex and complicated to understand by the consumer itself. The consumer also faces issues in the claim and settlement process (Shelkovnikov, 2016). Presently, insurance companies follow lengthy procedures for the initiation, maintenance, and closure of the policies. This resulted in a slow transaction process and payment settlement (Raikwar et al., 2018). A third-party involvement is required in the claim subrogation, making the process time-consuming and costly (IBM, 2018a). Blockchain technology provides various benefits for the insurance industry, including automation, reduction in fraud, and transparency. It helps increase the speed of initiating the contract and claim settlement. Marine insurance, travel insurance, policy underwriting, and peer-to-peer insurance are some of the blockchain applications in the insurance industry (Brophy, 2020).

3.3.9 Other applications

Blockchain can serve as a remedy tool in the other applications of the banking and finance services, including asset management, digital identity verification, loan syndication, consortium banking, and supply chain management. Blockchain technology records the data in a decentralised distributed ledger which replaces the centralised intermediary system. This will attract the buy-side firms by improving the data and reporting accuracy, reduce delays, and reduce the cost in asset management (PWC, 2016). The adoption of blockchain technology in supply chain financing will improve the company's future profits and reduce the operation risk (Choi, 2020). Blockchain, initially created as an alternative to traditional financial services, is drawing significant attention from financial institutions and banks, considering the importance of its applications in improving the sector's existing procedures and operations.

3.4 CHALLENGES OF BLOCKCHAIN TECHNOLOGY

Since its inception in 2008, blockchain technology has made significant disruption to various industries. Owing to its features, such as immutability and decentralisation, the technology outlines a promising solution for most pain points existing in the financial system. However, despite such potential of this innovation, there are many setbacks and hurdles in its implementation, as discussed below.

3.4.1 Scalability

With the growing acceptance of blockchain technology, the issue of scalability has become more evident (Garg et al., 2021). Since blockchain is a distributed ledger, its cumulative processing capability precisely relies on the estimated power of the instrument involved. As more and more people and nodes associate with the network, there are high chances of slowing down the entire network. Compared to Visa's 1,700/second, blockchain technology can only deal with an average of 4.6 transactions/second (Kot, 2019). This difference indicates a huge challenge for blockchain to implement globally in the banking and financial sector, where millions of transactions are required to be processed in a short time.

3.4.2 Security

The blockchain network participants are required to use public and private cryptographic keys to validate and verify their transactions (Garg et al., 2021). If under any circumstances, the user lost or forgot their private key or the device with the blockchain access is stolen or damaged, the system would not be able to recover the information or protect the data. Also, blockchain-based structures are susceptible and vulnerable to security frauds and cyber-attacks. The security of blockchain can be compromised through different measures such as 51% attack, double spending, distributed-denial-of-service (DDoS) attack and cryptographic cracking. Since 2009, the blockchain-based platforms, namely Ethernet and Bitcoin, have been embezzled successfully, resulting in a loss of 600 million yuan (Chang et al., 2020).

3.4.3 High energy consumption

The enormous amount of energy that blockchain technology use is what makes it not so ideal for the real world. Blockchain technology requires a high level of connectivity and significant energy consumption to operate on a large scale (Gan et al., 2021). Indeed, a single Bitcoin transaction requires a terawatt-hour amount of electricity (Chang et al., 2020). The proof-of-work consensus algorithm used by blockchain depends on miners to deal with complicated mathematical problems. Every time a transaction is updated on the ledger, there is considerable electricity consumption in the mining process. As reported by Blockchain Luxembourg, the cost of a single transaction of Bitcoin bounces around $75–$160, where electricity costs cover a significant chunk (Bloomberg, 2018).

3.4.4 Regulation and governance

There are no policies and legal frameworks governing the blockchain technology spectrum. The novelty and immaturity of blockchain applications make it even more challenging and complex to undertake its regulations in the financial sector (Osmani et al., 2021). On top of that, the difference between jurisdictions in different countries has increased the difficulty to create a universal law. The absence of these rules has hampered the application of this technology in various industries. Banks and financial institutions are concerned about their liability in the situation of bankruptcy, fraud, loss of consumer funds, or any other failures since there are no provisions for compensations and settlements. The situation becomes more questionable for organisations working in multiple jurisdictions.

3.4.5 Cost and efficiency

Though blockchain allows significant savings in time and transaction costs, the initial high cost of setting up the infrastructure is a major setback in its implementation (Reyna et al., 2018). There are various costs associated with its implementation, including the cost of designing, the cost related to hiring and managing developers, the license fee, etc. Small financial companies and banks may not be able to provide such high investment and recurring payments for storing data indefinitely, thus, hindering the adoption of blockchain on a large scale.

3.5 GLOBAL SCENARIO OF BLOCKCHAIN TECHNOLOGY IN BANKING AND FINANCIAL SERVICES

Blockchain technology is gaining the attention of the major world players in the banking and financial services industry due to its use cases, features, benefits, and possible solutions to the existing challenges in the traditional system. Presently, these banks are indulged in various blockchain-based consortiums to provide solutions and facilitate the services in the major applications of blockchain technology for the banking and financial services industry. Some of the major projects and consortiums are highlighted in Table 3.1.

Table 3.1 Global scenario of blockchain technology

Project/Platform/Company/ Consortium	Description
Indian Banks' Blockchain infrastructure Co Pvt. Ltd (IBBIC) *Launch Year: 2021* (G, 2021)	The 15 banks have created a consortium and established a company called IBBIC to facilitate and provide services for the inland letter of credit using blockchain technology. The primary purpose of the consortium is to improve efficiency and reduce transaction time. The consortium includes State Bank of India, HDFC Bank, ICICI Bank, Axis Bank, Kotak Mahindra Bank, Canara Bank, Yes Bank, Bank of Baroda, Indian Bank, IDFC first bank, Federal Bank, Induslnd Bank, South Indian Bank, and Standard Chartered.
Smart Dubai (Dubai Pay) *Launch Year: 2021* ("Dubai Blockchain Strategy," 2021)	Under Smart Dubai, DubaiPay is a shared payment service that provides a 24/7 payment facility by the government to their customers. This platform allows payment services in an easy-to-use, real-time, faster, and secure manner. The government is using this platform for real-time settlement and reconciliation.
Confirm *Launch Year: 2021* (Morgan, 2021)	Using blockchain technology, J.P. Morgan has provided a global account information validation application called confirm on the Onyx platform (by J.P. Morgan). It will help banking and financial institutions by improving the payment system. Using confirm, the cost would be significantly reduced for both the parties (banks)-sender and receiver, enabling a single source of truth and providing immutable records that reduce frauds in the payment system.

(Continued)

Table 3.1 (Continued) Global scenario of blockchain technology

Project/Platform/Company/ Consortium	Description
Contour *Launch Year: 2020* (Contour, 2020)	Contour is a global network of partners, including trade partners, corporates, and banks. This network aims to revolutionise the trade finance industry by building a decentralised network led by trust, collaboration, and innovation. This network will improve efficiency and reduce cost and complexity. Contour used R3's Corda Platform. Nine global banks back contour: Bangkok Bank, BNP Paribas, Citi, CTBC Bank, HSBC, ING, SEB, Sumitomo Mitsui Financial Group (SMBC), and Standard Chartered.
Thailand VAT Refund for Tourists (VRT) *Launch Year: 2020* (The Revenue Department of Thailand, 2020)	The Revenue Department of Thailand, along with alliance partners: Krungthai Bank public company Ltd, the immigration bureau, and the Customs department, have developed an application called Thailand VRT based on blockchain technology for providing a VAT refund system to the world's tourists.
UAE KYC Blockchain Consortium *Launch Year: 2020* (Insights, 2020)	The KYC Blockchain consortium was launched in 2020 by the Department of Economic Development (Dubai) with banks including HSBC, Abu Dhabi Commercial Bank, RAKBANK, Emirates Islamic, and Mashreq Bank. This platform provides users the facility of opening instant bank account without passing through the lengthy process of physical KYC validation. It will help optimise the cost, reduce the time for processing, and improve the business efficiency in the work processes. The consortium is using Norbloc's Flides blockchain platform.
Komgo *Launch Year: 2019* (Consensys, 2019)	Komgo is the first world platform based on blockchain technology to provide the ecosystem in commodity trade. Komgo delivers a highly secure and efficient way to transact digitally for the various stakeholders, including banks, carriers, traders, and other authorised partners. In just 1 year, Komgo has crossed 1 billion USD financing.
Global Financial Innovation Network (GFIN) *Launch Year: 2019* (GFIN, 2019)	An international group of financial regulators launched the GFIN in 2019. More than 70 organisations support the network. The network would help advance financial services by promoting financial inclusion, strengthening financial stability, and ensuring consumer wellbeing and protection.
We.Trade *Launch Year: 2018* (IBM, 2018b)	We.Trade is a platform in which 12 European banks start with two technology providers. The shareholder banks are Caixa Bank, Deutsche Bank, Erste Group, HSBC, KBC, Nordea, Rabobank, Santander, Société Générale, UBS, UniCredit, and CRIF, which used the IBM blockchain for facilitating and simplifying the international trade.

(Continued)

Table 3.1 (Continued) Global scenario of blockchain technology

Project/Platform/Company/ Consortium	Description
openIDL Launch Year: 2018 (AAIS, 2018)	openIDL is a distributed and open-source platform that is built on blockchain technology. The platform brings high security, efficiency, and generates transparency in regulatory reporting for the insurance companies. OpenIDL was developed and launched by the American Association of Insurance Services.
Global Blockchain Bond (Bond-*i*) by World Bank Launch Year: 2018 (IBRD, 2018)	World Bank has launched the first blockchain-based Bond-*i*. The arranger for the bond is the Commonwealth Bank of Australia. It would help the capital market be more secure, faster, and more efficient.
Project Ubin Launch Year: 2016 (MAS, 2020)	Project Ubin is a multi-year and multi-phase project that uses distributed ledger technology (DLT) to facilitate the clearing and settlements of securities and payments. The project has completed its 5-year journey and phase 5, resulting in support for various use cases, including delivery-versus-payment, inter-bank payments, cross-border payments, and payment commitments for trade finance.
Nonghyup Bank (Digital Identities) (Fitzpatrick, 2020)	Nonghyup bank has rolled out a digital identity service based on blockchain technology. The mobile employee Id is used to tighten the security of the personal data of the employees. The mobile identity is backed by several members, including LG, Hana Financial Group, and Samsung.
B3i Launch Year: 2016 (B3i, 2022)	B3i is a global initiative founded in 2016 as an insurance industry consortium. Presently, it is owned by 21 insurance market participants across the globe. The primary purpose of the Consortium is to provide solutions for reinsurance by using blockchain technology. It would provide customers a faster and cost-effective access to services.
Tradle (Tradle, 2017)	Tradle is building a network for providing secured and trusted know your customer requirements using blockchain technology. It will help lower the cost of know your customer work and increase the speed of operations processing.
Blockchain platform for the distribution of investment funds (Deloitte, 2022)	Deloitte Portugal is developing a platform using blockchain technology to distribute investment funds. The platform would facilitate the distribution of funds in Portugal in a more efficient way by reducing the involvement of the third party. Deloitte Portugal implements the proof-of-concept along with the EMEA FSI grid blockchain lab.

Apart from the above projects/use cases/consortium, Central Banks from various countries are focusing on developing and implementing the CBDC based on blockchain or distributed ledger technology (DLT). According to the Atlantic Council (2022), currently, 90 countries are exploring the CBDC. Some of them have been launched and implemented for retail or wholesale use. Some CBDC projects by various central banks are Sand Dollar (Bahamas), Digital Yuan (China), Digital Euro (Eurozone), Digital Yen (Japan), Digital Won (South Korea), Digital Pound (United Kingdom), e-Peso (Uruguay), Digital Dollar (United States), e-Hryvnia (Ukraine), Digital Baht (Thailand), Digital Ruble (Russian Federation), e-Krona (Sweden), and Dinero Electrónico (Ecuador) (Decentralized Dog, 2021).

3.6 CONCLUSION

Since the inception of blockchain technology, it has transformed the way businesses operate. It's not just information technology but also institutional technology (Davidson et al., 2018). Blockchain technology provides a decentralised economic infrastructure that helps lower the cost of institutional innovation in the organisation (Allen et al., 2020). To understand the potential of blockchain technology in banking and financial services, we have outlined its advantages and the overall development of this technology in different banking and financial services applications. The world is advocating a rising adoption of blockchain technology in this industry. Many banks and financial institutions have recognised its capabilities and are inclined to implement blockchain technology in their daily operations.

Despite such benefits, blockchain is still immature and needs to be closely scrutinised before implementing in the financial sector. The risk and challenges include scalability, initial cost, security, high energy consumption, regulation, and governance. These challenges could be a hurdle to the successful adoption of blockchain technology. There is a need to explore different approaches to overcome these challenges. Scalability is the most challenging issue as the Bitcoin platform can process only seven transactions/second; in contrast, there is a requirement to process millions of transactions in a short time. So, it is essential to build other platforms using blockchain technology that can handle and process millions of transactions to make this system efficient. Presently, various outlets of blockchain technology require high computational power, which is a serious matter of concern. Though renewable energy could be a solution to power production (Ghosh & Das, 2020), there is a need to reduce the energy consumption of blockchain technology. To address security concerns, several perspectives should be considered for making the system resilient, including smart contract security, associated surround system security, transaction-level security, network-level security, and ledger-level security (IDRBT, 2017). Also, banks, financial institutions, and government can collaborate to make policies for optimal use and successful implementation of blockchain technology as released by the Group of Seven (G7) named "public policy principles for retail central bank digital currencies".

REFERENCES

AAIS. (2018). *It's Time to Rethink Regulatory Reporting.* OpenIDL is here. Retrieved from https://aaisonline.com/openidl#:~:text=openIDL.

Ali, O., Ally, M., Clutterbuck, & Dwivedi, Y. (2020). The state of play of blockchain technology in the financial services sector: A systematic literature review. *International Journal of Information Management, 54,* 102199. https://doi.org/10.1016/j.ijinfomgt.2020.102199.

Allen, D. W. E., Berg, C., Markey-Towler, B., Novak, M., & Potts, J. (2020). Blockchain and the evolution of institutional technologies: Implications for innovation policy. *Research Policy, 49*(1), 103865. https://doi.org/10.1016/j.respol.2019.103865.

Atlantic Council. (2022). *Central Bank Digital Currency Tracker.* Retrieved from https://www.atlanticcouncil.org/cbdctracker/

B3i. (2022). *Driving Change in Insurance.* Retrieved from https://b3i.tech/.

Bloomberg, J. (2018). *Don't Let Blockchain Cost Savings Hype Fool You.* Retrieved from https://www.forbes.com/sites/jasonbloomberg/2018/02/24/dont-let-blockchain-cost-savings-hype-fool-you/?sh=1205b1285811

Brophy, R. (2020). Blockchain and insurance: A review for operations and regulation. *Journal of Financial Regulation and Compliance, 28*(2), 215–234. https://doi.org/10.1108/JFRC-09-2018-0127.

Buehler, K., Chiarella, D., Heidegger, H., Lemerle, M., Lal, A., & Moon, J. (2015). *Beyond the Hype: Blockchain in Capital Markets.* Retrieved from https://www.mckinsey.com/industries/financial-services/our-insights/beyond-the-hype-blockchains-in-capital-markets

Chang, V., Baudier, P., Zhang, H., Xu, Q., Zhang, J., & Arami, M. (2020). How Blockchain can impact financial services – The overview, challenges and recommendations from expert interviewees. *Technological Forecasting and Social Change, 158,* 120166. https://doi.org/10.1016/j.techfore.2020.120166.

Choi, T. M. (2020). Supply chain financing using blockchain: Impacts on supply chains selling fashionable products. *Annals of Operations Research.* https://doi.org/10.1007/s10479-020-03615-7.

CoinMarketCap. (2021). *Cryptocurrency Prices, Charts and Market Capitalizations.* Retrieved from https://coinmarketcap.com/

Consensys. (2019). *Komgo: Blockchain Case Study for Commodity Trade Finance.* Retrieved from https://consensys.net/blockchain-use-cases/finance/komgo/.

Contour. (2020). *About Contour: Leading the Digitalisation of Trade Finance.* Retrieved from https://contour.network/about-contour/.

Daluwathumullagamage, D. J., & Sims, A. (2021). Fantastic beasts: Blockchain based banking. *Journal of Risk and Financial Management, 14*(4), 170. https://doi.org/10.3390/jrfm14040170.

Davidson, S., De Filippi, P., & Potts, J. (2018). Blockchains and the economic institutions of capitalism. *Journal of Institutional Economics*, 14(4), 639–658. https://doi.org/10.1017/S1744137417000200.

Deloitte. (2015). *How Blockchain Can Reshape Trade Finance Pain Points*. Retrieved from https://www2.deloitte.com/content/dam/Deloitte/global/Documents/grid/trade-finance-placemat.pdf

Deloitte. (2017). *Blockchain in Banking While the Interest Is Huge, Challenges Remain for Large Scale Adoption*. Deloitte.

Deloitte. (2022). *Deloitte Creates a Blockchain Platform for the Distribution of Investment Funds*. Retrieved from https://www2.deloitte.com/pt/en/pages/financial-services/articles/blockchain-press-release.html.

Du, W. (Derek), Pan, S. L., Leidner, D. E., & Ying, W. (2019). Affordances, experimentation and actualization of FinTech: A blockchain implementation study. *Journal of Strategic Information Systems*, 28(1), 50–65. https://doi.org/10.1016/j.jsis.2018.10.002.

Dubai Blockchain Strategy. (2021). Retrieved from Digital Dubai website: https://www.digitaldubai.ae/initiatives/blockchain.

Fitzpatrick, L. (2020). *South Korea: Nonghyup Bank Introduces Digital Identities*. Retrieved from Forbes website: https://www.forbes.com/sites/lukefitzpatrick/2020/03/09/south-korea-nonghyup-bank-introduces-digital-identities/?sh=24993d947111.

G, J. (2021). *IBBIC: Paving Way for Blockchain Adoption by Indian Banks*. Retrieved from https://www.grantthornton.in/insights/blogs/ibbic-paving-way-for-blockchain-adoption-by-indian-banks/.

Gan, Q. Q., Lau, R. Y. K., & Hong, J. (2021). A critical review of blockchain applications to banking and finance: A qualitative thematic analysis approach. *Technology Analysis and Strategic Management*, 1–17. https://doi.org/10.1080/09537325.2021.1979509.

Garg, P., Gupta, B., Chauhan, A. K., Sivarajah, U., Gupta, S., & Modgil, S. (2021). Measuring the perceived benefits of implementing blockchain technology in the banking sector. *Technological Forecasting and Social Change*, 163, 120407. https://doi.org/10.1016/j.techfore.2020.120407.

GFIN. (2019). *The GFIN Is the International Network of Financial Regulators and Related Organisations Committed to Supporting Financial Innovation in the Best Interests of Consumers*. Retrieved from https://www.thegfin.com/.

Grigg, I. (2005). *Triple Entry Accounting*. Systemics Inc, 1–10.

Guo, Y., & Liang, C. (2016). Blockchain application and outlook in the banking industry. *Financial Innovation*, 2(1). https://doi.org/10.1186/s40854-016-0034-9.

Gupta, A., & Gupta, S. (2018). Blockchain technology: Application in Indian Banking Sector. *Delhi Business Review*, 19(2), 75–84.

IBM. (2018a). *Transforming Insurance Management with IBM Blockchain*. Retrieved from https://www.ibm.com/downloads/cas/OMJRXZAL

IBM. (2018b). *We.Trade*. Retrieved from https://www.ibm.com/case-studies/we-trade-blockchain.

IBRD. (2018). *World Bank Prices First Global Blockchain Bond, Raising A$110 Million.* Retrieved from The World Bank website: https://www.worldbank.org/en/news/press-release/2018/08/23/world-bank-prices-first-global-blockchain-bond-raising-a110-million.

IDRBT. (2017). *Application of Blockchain Technology to Banking and Financial Sector in India.* Institute for Development and Research in Banking Technology (Established by Reserve Bank of India).

Insights, L. (2020). *Duabi Launches Blockchain Know Your Customer Consortium.* Retrieved from https://www.ledgerinsights.com/dubai-blockchain-kyc-know-your-customer-consortium/

Jagtap, B., Sinnarkar, M., & Baul, S. (2019). *Blockchain in BFSI Market Statistics, Segments Analysis: Forecast–2026.* Retrieved from https://www.allied-marketresearch.com/blockchain-in-bfsi-market#:~:text=Blockchain%20Market%20In%20BFSI%20Sector,73.8%25%20from%202019%20to%202026

Kabir, M. R. (2021). Behavioural intention to adopt blockchain for a transparent and effective taxing system. *Journal of Global Operations and Strategic Sourcing, 14*(1), 170–201. https://doi.org/10.1108/JGOSS-08-2020-0050.

Kot, I. (2019). *3 Major Raodblocks to Blockchain Adoption in Banking.* Retrieved from https://www.finextra.com/blogposting/18197/3-major-roadblocks-to-blockchain-adoption-in-banking#:~:text=Currently%2C%20three%20major%20blockchain%20limitations,scalability%2C%20and%20regulatory%20compliance%20risks

Lee, I., & Shin, Y. J. (2018). Fintech: Ecosystem, business models, investment decisions, and challenges. *Business Horizons, 61*(1), 35–46. https://doi.org/10.1016/j.bushor.2017.09.003.

MAS. (2020). *Project Ubin: Central Bank Digital Money using Distributed Ledger Technology.* Retrieved from https://www.mas.gov.sg/schemes-and-initiatives/Project-Ubin.

Morgan, J. P. (2021). *Validate Account Information Globally with Speed, Simplicity, and Security.* Retrieved from ONYX website: https://www.jpmorgan.com/onyx/confirm.htm.

Nakamoto, S. (2008). *Bitcoin: A Peer-to-Peer Electronic Cash System.* Decentralized Business Review.

Noble, D., & Patil, K. (2021). Blockchain in stock market transformation: A systematic literature review. *Revista Gestão Inovação e Tecnologias, 11*(4), 5088–5111. https://doi.org/10.47059/revistageintec.v11i4.2551.

Osmani, M., El-Haddadeh, R., Hindi, N., Janssen, M., & Weerakkody, V. (2021). Blockchain for next generation services in banking and finance: Cost, benefit, risk and opportunity analysis. *Journal of Enterprise Information Management, 34*(3), 884–899. https://doi.org/10.1108/JEIM-02-2020-0044.

Patki, A., & Sople, V. (2020). Indian banking sector: Blockchain implementation, challenges and way forward. *Journal of Banking and Financial Technology, 4*(1), 65–73. https://doi.org/10.1007/s42786-020-00019-w.

Persson, T., Parker, C. F., & Widmalm, S. (2017). Social trust, impartial administration and public confidence in eu crisis management institutions. *Public Administration*, 95(1), 97–114. https://doi.org/10.1111/padm.12295.

PWC. (2016). *Blockchain in Asset Management*. Retrieved from https://www.pwc.co.uk/financial-services/fintech/assets/blockchain-in-asset-management.pdf

Qiu, T., Zhang, R., & Gao, Y. (2019). Ripple vs. SWIFT: Transforming cross border remittance using blockchain technology. *Procedia Computer Science*, 147, 428–434. https://doi.org/10.1016/j.procs.2019.01.260.

Raikwar, M., Mazumdar, S., Ruj, S., Gupta, S., Chattopadhyay, A., & Lam, K.-Y. (2018). A blockchain framework for insurance processes. In *2018 9th IFIP Internatiional Conference on New Technologies, Mobility and Security (NTMS)*, 1–4. IEEE.

Rana, K. (2020). *Triple Entry Accounting System: A Revolution with Blockchain*. Retrieved from https://medium.com/dataseries/triple-entry-accounting-system-a-revolution-with-blockchain-768f4d8cabd8

RBI. (2010). *Master Circular – Know Your Customer (KYC) Norms/Anti-Money Laundering (AML) standards/Combating of Financing of Terrorism (CFT)/ Obligation of Banks under PMLA, 2002*.

Reyna, A., Martín, C., Chen, J., Soler, E., & Díaz, M. (2018). On blockchain and its integration with IoT. Challenges and opportunities. *Future Generation Computer Systems*, 88, 173–190. https://doi.org/10.1016/j.future.2018.05.046.

Sandeep, K. V. (2021). *Blockchain in Capital Markets*. Retrieved from https://www.wipro.com/capital-markets/blockchain-in-capital-markets/

Shelkovnikov, A. (2016). Blockchain applications in insurance. *Deloitte Report*, 1–2.

Thailand, T. R. D. (2020). *The Revenue Department launches "The World's First Blockchain VAT Refund Technology."* Retrieved from http://vrtweb.rd.go.th/index.php/en/component/vrt/main/286?layout=detail.

Tradle. (2017). *KYC on Blockchain*. Retrieved from https://tradle.io/.

World Bank. (2018). *Financial Inclusion*. Retrieved from https://www.worldbank.org/en/topic/financialinclusion

Yu, T., Lin, Z., & Tang, Q. (2018). Blockchain: The introduction and its application in financial accounting. *Journal of Corporate Accounting & Finance*, 29(4), 37–47. https://doi.org/10.1002/jcaf.22365.

Chapter 4

Safeguarding blockchains from adversarial tactics

Alex Neelankavil Devassy and Dhanith Krishna
Ernst & Young GDS

CONTENTS

4.1 INTRODUCTION: BACKGROUND AND DRIVING FORCES

Blockchain platforms and applications will play a pivotal role in shaping Industry 4.0. The main reason being industry 4.0 relies on the convergence of cyber-physical systems and Internet of Things (IoT) where

DOI: 10.1201/9781003282914-4

edge devices gain more prominence as compute nodes and should support massive machine-to-machine communication with no human interaction. Blockchain is the ideal technology to empower such autonomous interconnected networks, due to decentralization of trust, built-in traceability and tamper resistance. Gartner predicts that business value addition through blockchain adoption could reach $176 billion by 2025 and $3.1 trillion by 2030 (Forecast: Blockchain Business Value, Worldwide, 2017–2030, 2022).

But most organizations that adopt blockchain technology to deploy their applications perceive blockchain as a silver bullet to threats from hackers and adversaries. The most popular rationale behind this misconception is because "Blockchains being branded to be secure from hackers because of its entire concept was built upon cryptography and immutability (i.e., information can be permanently stored on a ledger without being tampered with)" (Anita and Vijayalakshmi, 2019). Another popular delusion on blockchain solutions is regarding data storage and privacy (Ehab et al., 2020), i.e., organizations may fail to ensure data privacy requirements for on-chain data which could result in data leakages. Market hype and lack of awareness push organizations to design use cases that are not compatible with blockchain or are counterproductive. Often key areas such as privacy and data security take a back seat in this arms race. Misinterpretation of the underlying technology, along with bad security practices, could create serious security vulnerabilities in blockchain applications. It is key for organizations to have a solid strategy for blockchain adoption by evaluating the suitability of different blockchain architectures, scalability, in-built security features in blockchain platforms and most importantly understanding adversarial attack vectors. Organizations should bake-in security across the blockchain adoption lifecycle by infusing security processes across plan, design, code, test, deploy and monitoring stages.

The objective of this chapter is not to establish that the underlying blockchain technology is fundamentally flawed but to highlight the security risks in blockchain applications and to guide secure blockchain implementation. Security vulnerabilities caused due to insecure coding practices, patch management issues and architectural flaws are discussed in detail along with mitigations. Since the chapter consists of technical deep dive with detailed case studies and proof of concepts, the scope of discussion is limited to Ethereum & Hyperledger Fabric, as two of the most popular blockchain platforms currently in the market. Another rationale behind choosing these platforms is to highlight the difference between adversarial tactics and techniques when attacking public and private blockchain solutions.

The chapter begins with Section 4.1, investigating the need to secure blockchain and debunking common myths. We introduce a four-layer model to study and evaluate the attack surface and associated attack vectors in Section 4.2. Using case studies, adversarial tactics are explained across the network layer (Section 4.3), smart contract (Section 4.4), access layer (Section 4.5) and application layer (Section 4.6), along with a discussion on mitigation steps. Further, the importance of adopting secure software development lifecycle (SSDLC) practices across various phases of development is established in Section 4.7. Concepts presented in this chapter are meant to help blockchain developers and architects to avoid security issues and design flaws, to develop blockchain solutions that are inherently resilient to cyber-attacks.

4.2 NECESSITY OF SAFEGUARDING DAPPS FROM ADVERSARY TACTICS

Decentralized apps (Dapps) are blockchain-native applications with smart contract-powered logic. Dapps are gaining prominence, particularly in finance, in the form of decentralized finance (De-Fi) use cases. By establishing cryptographic trust, Dapps eliminate the need for intermediaries and supports trust-free transactions. If all participants agree on the ground rules established through smart contracts, Dapps can ensure compliance with the underlying contract logic without an intermediary or owners.

Depending on the nature of the use case, Dapps can be deployed on private or public blockchains. When Dapps are used to solve a public use case, they are deployed onto a public blockchain. Whereas when the participants are limited to a small set of organizations, consortium or federated blockchains are used to host Dapps. Finally, if Dapps are used to solve particular use cases within an organization, they can be hosted on a blockchain which is fully owned by one organization. For the sake of simplicity, we are referring to both consortium and private blockchains as private blockchain in this chapter (Figure 4.1).

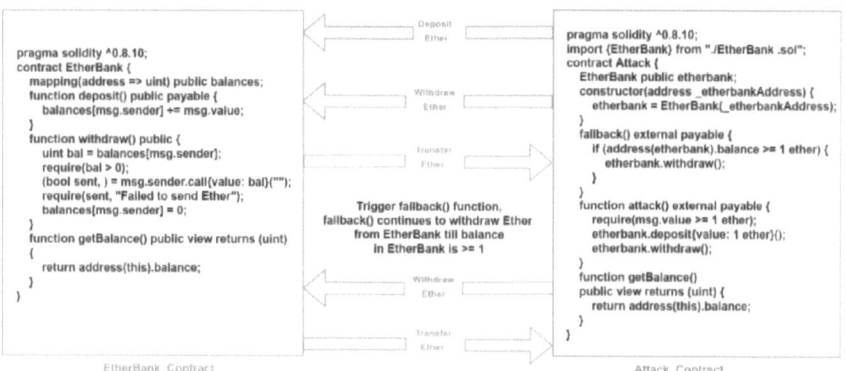

Figure 4.1 EtherBank contract (vulnerable to reentrancy attack), attack contract (deployed for exploiting EtherBank).

Even though Dapps inherit the cryptographic trust and immutability of the underlying blockchain platform, they are far from being resilient to cyber-attacks. Dapps can be vulnerable to security vulnerabilities like any other software. The attack surface of Dapps varies widely based on blockchain platform, type of blockchain and the tech-stack of the application.

4.3 BLOCKCHAIN REFERENCE MODEL

In order to study the attack surface of Dapps, we need to inspect various layers in a blockchain solution and understand potential attacks on each layer. Most blockchain-based Dapps contain below layers in their architecture:

1. **Network Layer:** Network or node layer of the blockchain consists of computers running blockchain software hosting a distributed ledger. Depending on the solution architecture, it is fairly common to find oracles and off-line storages at this layer.
2. **Smart Contract Layer:** Smart contract layer consists of smart contracts and other chaincode that form the software layer of a blockchain.
3. **Access Layer:** Access layer is a middleware layer that provides connectivity between smart contracts and end-user applications. Depending on the blockchain technology, access layer can consist of Web2 Application Programming Interfaces (APIs) or Remote Procedure Call (RPC) clients. However, it is not mandatory to have an intermediate layer as end-user applications can use Web3 calls to interact directly with the smart contracts.
4. **Application Layer:** End-user applications are typically web or mobile applications that are used by customers to perform operations on the blockchain solution. Organizations that develop blockchain solutions should be aware of how adversaries exploit well-known attack vectors originating from end-user applications to compromise blockchain solutions.

Depending on the blockchain platform and type of platform, the attack surface of each of these layers varies. Generally, as we move from the network to the application layer, the attack surface increases.

4.4 NETWORK LAYER ATTACKS IN BLOCKCHAIN

When it comes to blockchain security, solution architects focus more on smart contract security or application layer security. Depending on whether the blockchain is public or private, sufficient attention should be given in securing the blockchain network and participating nodes. This is particularly important for private blockchains due to the relatively small number of nodes.

In Ethereum public blockchain, production-grade dapps are deployed on the Ethereum Mainnet, while most development takes place on Testnets like Rinkeby and Ropsten (Networks | ethereum.org 2022). Similar to how cloud providers relieve their users from infrastructure security implementations, these predeployed networks relieve solution creators from worrying about Ethereum node security. Despite Ethereum's Mainnet and Testnets, developers who need faster transaction speeds opt to build their own private Ethereum network, and academic institutions may prefer their consortium of networks (Networks|ethereum.org 2022) to Ethereum's Mainnet. In this case, Ethereum's nodes or network security must be overseen by developers or institutions.

However, in Hyperledger Fabric in spite of the detailed guidelines provided by Fabric, if node configuration is not reviewed from a security angle, security vulnerabilities may occur. This is primarily due to the fact that developers tend to use default configuration settings without understanding security implications. To breach the targeted system, advisories frequently rely on the most basic of developer errors.

Below sections cover some of the network layer attack vectors; adversaries could leverage to compromise the targeted blockchain network. Attack vectors discussed here include tactics used by adversaries to locate vulnerable blockchain nodes, compromising the integrity of blockchain networks via attacks on consensus and identity management mechanisms.

4.4.1 Node configuration flaws

Secure node configuration is fundamental to ensuring network security in blockchain networks. A poorly configured private Ethereum network could expose its services via JSON RPC endpoints. It's fairly straightforward for adversaries to take advantage of this security flow and grant themselves access to the network. Figure 4.2 demonstrates the simplified Nodejs code in which attackers could establish a connection with a vulnerable Ethereum node.

An attacker who can access the network of nodes could perform various fraudulent activities including user enumeration, password brute force, etc. A more sophisticated adversary would remain silent in the network to passively monitor smart contract code and associated workflow in order to exploit the bugs, while the same smart contract is deployed to Mainnet, where ether has its value.

Using default credentials is one of the most popular methods advisory uses to gain access. In Hyperledger Fabric, a developer must utilize the bootstrap administrator's enrollment credentials to start the CA (Certificate Authority) server. To start the CA server with fabric-ca-server Command

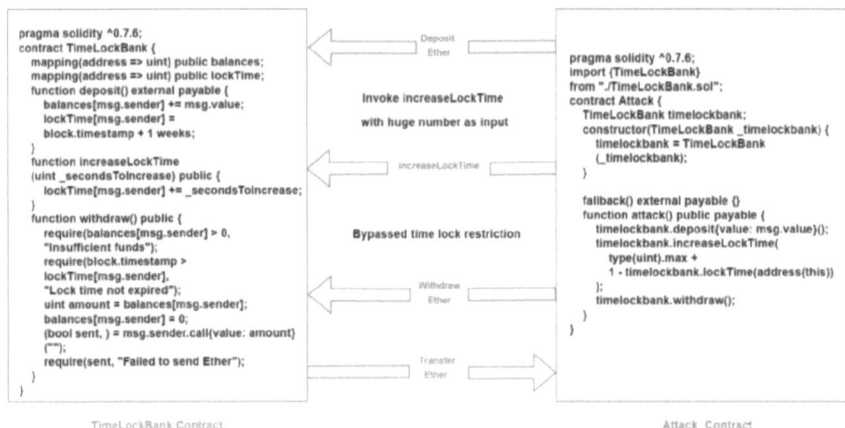

Figure 4.2 Simplified Nodejs code to establish a connection with a vulnerable Ethereum node.

Line Interface (CLI), Hyperledger Fabric recommends the command "fabric-ca-server start -b admin:adminpw" in its official documentation (Fabric CA User's Guide – hyperledger-fabric-cadocs main documentation 2022).

A developer using the above command with the default username "admin" and password as "adminpw" leaves the CA server (one of the critical components in Fabric architecture) vulnerable to various attack vectors such as attacker login with default enrollment credentials, brute force, password spraying, etc. Even though Fabric recommends not using default enrollment credentials, it does not force users to provide custom strong credentials during CA configuration.

Similar to bootstrapping CA server, developers often use or leave default settings as it is while configuring off-chain storage databases such as CouchDB. Hyperledger Fabric recommends running CouchDB container on the same server as the peer, for this developer needs to configure CouchDB settings in docker-compose-couch.yaml, which is later used to spin up CouchDB container. In Fabric's official GitHub repository, CouchDB credentials (Fabric-samples/compose-couch.yaml at main·hyperledger/fabric-samples 2022) are configured as "admin" and "adminpw". An attacker who has access to CouchDB endpoints can try accessing backend database and list available DBs using simple http query such as "http://admin:adminpw@"+<targetip>+":"+<targetport>+"/_all_dbs".

By impersonating an admin user, an attacker has admin privileges in a database, which would affect the confidentiality, integrity and availability of the whole solution and would even cause huge financial/reputation loss to the victim organization.

In Hyperledger Fabric, operations service endpoints should be restricted to authorized personal. Operations services are hosted by peer and orderer nodes for exposing APIs which are consumed by operators of Fabric network for multiple purposes such as to manage the active logging specification, to determine the liveness and health of nodes in network, for understanding behavior of the system and to query version of nodes. Attackers could use exposed operations service endpoint to enumerate backend system information through logs, operational metrics and other internal details.

Since operation service APIs are unrelated to Fabric network and are only consumed by operators, architects and developers of the network may neglect associated security best practices. A misconfigured operations service could leak out sensitive information which identifies the victim node as Hyperledger Fabric peer or orderer to an attacker who enumerates the network for possible Fabric nodes. For example, curl utility can be used by attackers to invoke vulnerable operations service endpoint for extracting version information of the fabric network using the command "curl http://targetip:targetport/version".

It is therefore essential for architects of the fabric network to ensure that operations services in Fabric are configured with adequately strong settings and access is restricted to operators. In its official documentation (The Operations Service, n.d.), Hyperledger Fabric recommends implementing mutual Transport Layer Security (TLS) with client certificate authentication on exposed APIs of the Operations service. Without enabling mutual TLS, organizations risk-exposing operations service endpoints to any unauthenticated attacker for exploitation.

4.4.2 Attacks on consensus

Sybil Attack: Resistance against sybil attack is a fundamental design consideration for blockchain platforms. Blockchain platforms achieve sybil resistance using consensus. In a Sybil attack, an attacker undermines a peer-to-peer network's trust model by creating a large number of pseudonymous identities and using them to wield disproportionately large influence in network (Iqbal & Matulevicius, 2021). Because blockchain systems operate on a peer-to-peer network, it is possible to run several phoney nodes. Furthermore, blockchain systems contain valuable digital assets that encourage attackers to carry out this assault. Once the attacker deployed sybil nodes gain recognition in the blockchain network, then the attacker could perform various fraudulent activities such as disrupting information flow, out-voting (or blocking) legit victim nodes, refusing to receive or transmit data, etc.

Figure 4.3 shows sample Hyperledger Fabric network with two channels and four participating organizations. Sample network has Org1, Org2 and Org3 as Channel 1 participants, while Org1 and Org4 are Channel 2 participants. Channel 1 enforces a MAJORITY policy for the endorsement of transactions in it. If Org2 is a malicious node in Channel 1, and if Org2 could compromise Channel 1 for obtaining control of the admin in the MSP

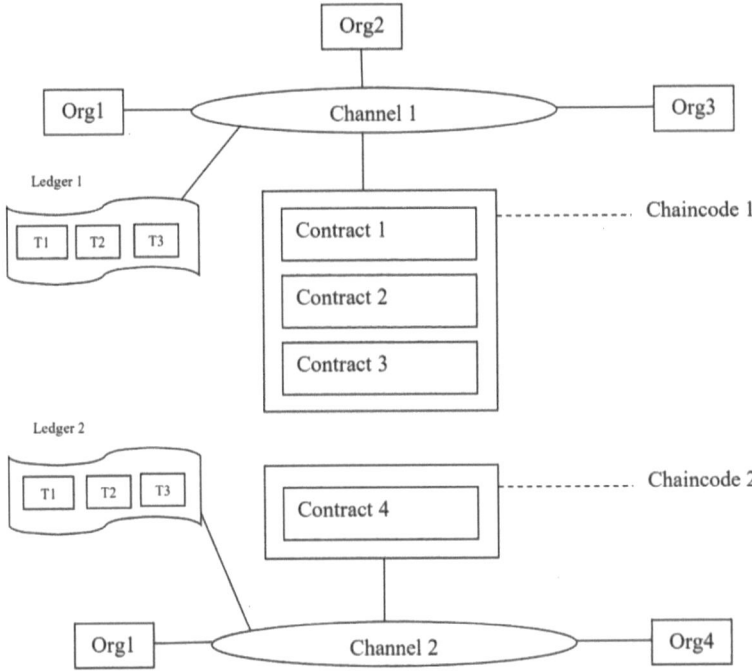

Figure 4.3 Sample fabric network with four organizations and two channels.

organization (governing organization), then it is possible for malicious node Org2 to perform sybil attack by spawning multiple nodes and joining them to Channel 1 such that Org2 could gain considerable influence on the network. This would allow Org2 to make fraudulent transactions by satisfying the endorsement policy (Dabholkar & Saraswat, 2019).

Double Spending Attack: The most widely used consensus algorithm in public blockchains such as Ethereum, Bitcoin, and others is proof-of-work (PoW). Each node in the PoW algorithm attempts to discover a nonce value that will yield a hash that matches a set of conditions. The complexity of computing such a nonce value can be controlled using the hash value's criteria. When a nonce value of this type is discovered, a block is created and published to the P2P network. Peer nodes pick the longest chain to publish updates. This allows the blockchain to continue growing while eliminating inconsistent blocks.

Mining is another name for this procedure of discovering nonce and publishing blocks to the chain. A peer node with higher processing power (also known as hashrate strength) can calculate the nonce value faster than a peer node with lower processing power and so has a greater chance of publishing new blocks into the blockchain network. This system, however, has a flaw. By triggering double spending, for example, a self-centered node or a group of collaborating nodes with aggregate hashrate strength greater than the combined hashrate strength of the remaining nodes in the network might damage the integrity of the blockchain system. A 51% attack is a term used to describe these types of attacks (Yang et al., 2019).

The feasibility of a 51% exploit appears to be difficult on large-scale public blockchains with a lot of participating nodes (like Bitcoin and Ethereum), but it is achievable on smaller blockchain systems which have inadequate amount of total computational power due to smaller number of participants (Sayeed & Marco-Gisbert, 2019). However, Ethereum Classic (ETC) is one of the public blockchain systems that have been hit by a 51% attack. ETC is the original version of Ethereum that survived after the core team forked the Ethereum network to reverse the infamous Decentralized Autonomous Organizations (DAO) hack (Sayeed et al., 2020). On January 5, 2019, it was reported that a total of 219,500 ETC worth $1.1 million were double spent in the attack (Iqbal & Matulevicius, 2021). However, exploitation of 51% attack is feasible on private blockchains due to small number of participating nodes, especially if there are existing security vulnerabilities in the nodes.

4.4.3 Identity and access management issues

Public blockchain networks use pseudo-anonymous identities for their participants. However, this is not suitable for most private blockchain solutions. Organizations running private blockchain may require that all participants are well known, and the transactions performed by all parties are traceable. Most blockchain platforms leverage digital certificates as the primary method to identity participants and make the network permissioned. CA servers perform a key role in issuing, maintaining and disposing of identities in most private blockchain platforms. Consequently, permissioned networks are generally very resilient to sybil attacks (Davenport et al., 2019).

However, a permissioned blockchain is only as secure as its CA infrastructure. CA infrastructure should be protected against attackers and malicious insiders. For instance, a malicious CA administrator would be able to add new identities to the blockchain network, adding existing identities to certificate revocation list (CRL) to block their access. Furthermore, identities issued by the CA are identified by the private keys associated. In that respect, secure storage and protection of the private keys are important to prevent identity theft. Another example of how improper implementation of CA leads to private key leakage can be observed in the adoption of particular CA settings in Hyperledger Fabric. Developers can choose how to manage a CA and create cryptographic materials with Hyperledger Fabric. FabricCA (an in-built mechanism in Hyperledger Fabric), Cryptogen and the use of third-party CA are all available options for developers to configure CA. The implementations of these CAs have problems of their own. Cryptogen generates all private keys in one centralized location, and it is up to the developer to copy them to the necessary hosts and containers in a timely and secure manner. As all private keys are stored in one location, the solution is susceptible to private key leakage attacks.

In a multiparty private blockchain (consortium blockchain), it will not be favorable for all participants to have access to all available data and contracts. Access control should be configured on data and smart contract access. For example, even though Hyperledger Fabric provides channels (Channels – hyperledger-fabricdocs main documentation 2022) and private data collections (PDCs) (Private data – hyperledger-fabricdocs main documentation 2022) to ensure data privacy by implementing strict access control on data, it may be still possible for an attacker to extract sensitive information from Fabric network with thorough enumeration.

Channels help to segregate the Fabric network into multiple subnetworks. Such that only participants of a subnetwork, i.e., channel can view/execute transactions on smart contracts deployed in that same channel. Figure 4.3 shows Org1, Org2 and Org3 as Channel 1 participants, while Org1 and Org4 are Channel 2 participants. In this particular setup, it is not possible for Org 4 to view/execute transactions that occur in Channel 1. Thus, channels in Hyperledger Fabric can offer privacy in certain scenarios. But what if Org1 and Org2 in Figure 4.3 are required to keep specific data private so that Org3 is unable to access it? PDCs in Fabric can be used in such scenarios when considerably greater fine-grained data privacy is required. Thus, PDCs can be leveraged when a subset of organizations in a channel needs to perform transactions which are shared among all participants of that channel together with ordering service nodes, but access to data within that transaction is only made accessible to that particular subset of organizations.

In Figure 4.3, a malicious user in Org3 can access Ledger 1 via Channel 1. In such a scenario, if the main part of Chaincode 1's proposal contains any private data pertaining to Org1 and Org2, it is possible for that malicious user in Org3 to view the same even though Org3 is not intended to access the private data of Org1 and Org2. To ensure that such data leaks do not occur in the channel, organizations can make use of a special field in the chaincode proposal called the transient field (Paulsen et al., 2021) because private data in the transient field is not included in a transaction, which is submitted to the ledger. Another scenario where a malicious user in Org3 can access restricted private data of Org1 and Org2 via Channel 1 is by leveraging brute force techniques on relatively simple and predictable private data. Such scenarios can be avoided if random salt is used in hashing predictable private data such that the resulting hash cannot be cracked via brute force.

4.4.4 Best practices

While deploying a private/consortium Ethereum, Hyperledger Fabric networks, solution architects or security engineers need to ensure that a proper network configuration is adopted including segregation of network using firewalls, ensuring secure node configurations and maintaining adequate patch levels. Such a proactive approach would enable organizations to shield against some of the level known network attack vectors. However, these controls may not be adequate to provide protection against attacks which involve zero-day exploits. For enhanced security coverage and increased attack resiliency, organizations should investigate reactive approaches, which would involve active monitoring of blockchain nodes, access logs and network traffic.

4.5 SMART CONTRACT ATTACKS IN BLOCKCHAIN

Smart contract security is a key concern when developing blockchain-based applications. Before getting into smart contract attacks, it's important to understand what "smart contract" means in Ethereum and what "smart contract"/"chaincode" means in Hyperledger Fabric. Below definition is from Ethereum's official documentation (Introduction to smart contracts|ethereum.org 2022).

> Smart Contract: A "smart contract" is simply a program that runs on the Ethereum blockchain. It's a collection of code (its functions) and data (its state) that resides at a specific address on the Ethereum blockchain.

Below definitions are from official Hyperledger Fabric documentation (smart contracts and chaincode – hyperledger-fabricdocs main documentation 2022).

> Smart Contract: defines the executable logic that generates new facts that are added to the ledger.
> Chaincode: A chaincode is typically used by administrators to group related smart contracts for deployment but can also be used for low level system programming of Fabric.

As illustrated in Figure 4.3, it is important to know that in Fabric, a smart contract is defined within a chaincode such that it's possible to define multiple smart contracts inside a chaincode (in Figure 4.3, contracts 1, 2 and 3 are defined in Chaincode 1). Below sections cover some of the attack vectors adversaries could leverage to manipulate the backend logic of blockchain-based applications. Attack vectors discussed here include tactics used by adversaries to exploit business logic, data validation and the deterministic characteristic in smart contracts.

4.5.1 Smart contract logic flaws

As more client-server apps are moved to the Ethereum's decentralized network, developers are forced to go through a steep learning curve. Without adequate blockchain security benchmarks in the industry and developer awareness, the security aspect of the developed blockchain applications is not getting sufficient attention. In Ethereum, adversaries often try to bypass or exploit smart contract logic because successful exploitation would enable an adversary to gain huge financial incentives in the form of crypto currencies. An infamous example of this can be observed in the DAO hack (Sayeed et al., 2020), where hackers exploited smart contract vulnerability known as reentrancy (Chinen et al., 2020) in order to transfer ether to a malicious account.

Reentrancy occurs when an external callee contract calls back to a function in the caller contract before the caller contract's execution is complete and the internal state of the contract is not updated before passing control to the external contract. Until the caller contract is drained of ether or the transaction runs out of gas, the attacker can get around the due validity check.

Reentrancy can be caused due to the below reasons

- A contract's control-flow decision relies on some of its state variable(s) that should be, but are not, updated by the contract itself before calling another (i.e., an external) contract.
- There is no gas limit when handing the control-flow to another contract.

Figure 4.1 illustrates the vulnerable smart contract EtherBank. Logic flow inside EtherBank is implemented such that withdraw() function inside the contract will transfer funds to msg.sender (callee) before it sets the balance to zero. An attacker can exploit withdraw() by recursively calling the function and draining the whole ether in EtherBank contract. Attacker calls attack() function in attack contract by providing one ether. Initially, attack() function deposits one ether to EtherBank and then calls withdraw() function of EtherBank contract which would transfer the correct amount of funds to the attacker. Once attack contract receives this amount, it triggers the fallback function inside it which calls withdraw() of EtherBank again and again till the balance of EtherBank is greater than or equal to 1. This would completely drain EtherBank transferring all its ether to attack contract.

The best practices (SWC-107·Overview 2022) to avoid reentrancy vulnerability includes

- Implementing Checks-Effects-Interaction's pattern in smart contracts which ensures that all internal state changes are performed before the call is executed
- Using reentrancy locks such as OpenZeppelin's ReentrancyGuard.

4.5.2 Insecure arithmetic

Insecure arithmetic vulnerability was first observed during the attack against the BeautyChain Coin (BEC) tokens (Gao et al., 2020). It occurs when the result of an arithmetic operation exceeds the range of a solidity data type, resulting in illegal manipulation of state variables, for example. The flaw is created by a solidity code that does not validate numeric inputs properly, and neither the solidity compiler nor the Ethereum Virtual Machine (EVM) enforces integer overflow/underflow detection.

Figure 4.4 illustrates vulnerable smart contract TimeLockBank. TimeLockBank primarily consists of three functions deposit (), increaseLockTime() and withdraw(). deposit() function of TimeLockBank is designed in such a way that a deposited amount can only be withdrawn after a week. However, it is possible for an attacker to leverage integer overflow issue in increaseLockTime() function of TimeLockBank contract. As the name suggests, increaseLockTime() function is used to increase the lock period on the deposited amount as per the user input. increaseLockTime() function had declared variable _secondsToIncrease of type unit256. A unit256 is an unsigned integer of 256 bits (unsigned, as in only positive integers). Since EVM is limited to 256 bits in size, the assigned number range is 0 to 4,294,967,295 (2^{256}). If we go over this range, the figure is reset to the bottom of the range ($2^{256}+1=0$). If we go under this range, the figure is reset to the top end of the range ($0-1=2^{256}$). Underflow takes place when we subtract a number greater than zero from zero, resulting in a newly assigned integer of 2^{256}.

Upon invoking attack() function inside attack contract, function deposits ether from attacker to TimeLockBank, next attack() function abuses integer

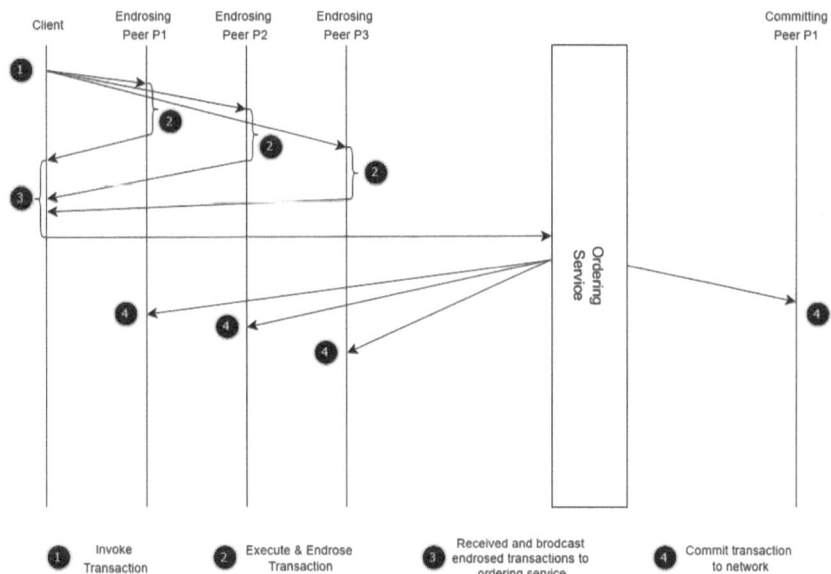

Figure 4.4 TimeLockBank contract (vulnerable to integer overflow), attack contract (deployed for exploiting TimeLockBank).

overflow issue in increaseLockTime() function by invoking increaseLock-Time() of TimeLockBank with a huge number as input. This would bypass time lock restriction and would enable the attacker to invoke withdraw() in TimeLockBank for withdrawing the deposited amount.

The best practices (SWC-101 · Overview 2022) to avoid integer overflow and underflow vulnerabilities include

- Using vetted safe math libraries for arithmetic operations consistently throughout the smart contracts.

4.5.3 Attacks on deterministic nature of smart contracts

Figure 4.5 depicts the transaction flow (Execute-Order-Validate) (Transaction Flow – hyperledger-fabricdocs main documentation 2022) in Hyperledger Fabric, which requires smart contracts to be deterministic (Sukhwani et al., 2018), i.e., when the same contract is deployed in multiple nodes and provided with the same input, they are expected to produce the same output. If a non-deterministic contract is invoked in the Fabric network, it would fail in the execute phase. This deterministic nature of smart contracts in Fabric restricts the use of a secure random number generator (RNG) in code. Because if a smart contract with RNG is deployed in multiple peers of Fabric, then invocation to that smart contract would result in every endorsing peer calculating a different random number, which does not comply with Fabric requirement that the smart contract needs to be deterministic. Developers often use a pseudorandom number generator (PRNG) as a workaround for this issue. A PRNG makes use of a deterministic algorithm to generate output. But the improper implementation of PRNG in smart contracts can compromise application security.

A most common example of improper implementation of PRNG in smart contract is the usage of getTxTimestamp() value, which would return the timestamp when the transaction was created as a seed to PRNG function. Imagine a smart contract "S" which makes use of getTxTimestamp() value as a seed to PRNG function is deployed on channel "C". Then, a malicious

```
var Web3 = require('web3')
let web3 = new Web3(new Web3.providers.HttpProvider("http://"+IP_address+":"+port));
web3.eth.net.isListening((e, r) => {
    if (r == true) {
        console.log("Established Connection With Ethereum Node")
    } //Closing if(r == true)
    else {
        console.log("Failed To Establish A Connection With Ethereum Node")
    }
```

Figure 4.5 Transaction flow in Hyperledger Fabric.

user in channel "C" who has access to both transaction timestamp and the smart contract "S" will be able to recompute the random number and thus compromise the security of smart contract logic.

It is possible to achieve secure implementation of PRNG in smart contract via various techniques. One being the practice of passing seed value for PRNG function as input data in the transient field of chaincode proposal (Paulsen et al., 2021). This would prevent participants of the channel from accessing the seed value since private data in the transient field is not included in the transaction which is submitted to the ledger. Other techniques involve the use of off-chain solutions like decentralized oracles such as Bitcoin Beacon (Jo and Park, 2019) to calculate random numbers. These solutions do not require to be deterministic in nature and can produce truly random numbers.

An adversary who tries to attack the business logic of applications deployed in Fabric might try to take advantage of vulnerabilities in smart contract such that attacker forces smart contract to be non-deterministic or behave unexpectedly. Improper use of global variables in smart contracts is one such vulnerability that is leveraged by attackers for forcing application to behave in an unexpected way. If a Fabric application's state-changing transaction depends on a vulnerable global variable "GV" in a smart contract "S", an attacker could put the application/smart contract out of service for legit users by sending a transaction proposal to only one peer in network which would update the value of global variable "GV" in smart contract "S". This is possible due to the fact that smart contract needs to be deployed in each peer who acts as participants of the corresponding channel and values stored in a global variable inside smart contract persist across transactions. By sending a transaction proposal which would update the value of global variable "GV" in smart contract "S" to only one peer in network, an attacker is changing/updating the state of global variable "GV" only on one peer, and all other peers in network retain the previous state of global variable "GV". Even though attacker's transaction fails to be committed in the ledger since endorsement policy requires the endorsement of all peers, the smart contracts "S" service will be denied to legit users because peers in the network have inconsistent values for the global variable "GV", which prevents them from arriving in a consensus.

It's recommended only to use global variables in smart contracts when it's absolutely necessary. Developers need to secure smart contracts by implementing proper error handling for tackling unexpected behaviors and for ensuring global variables in smart contracts are only modified in a valid transaction (Paulsen et al., 2021). It's also necessary to make sure that the smart contract logic flow is thoroughly tested for uncovering unexpected branches/flaws in code.

4.5.4 Best practices

Even while the language of smart contracts differs among blockchain platforms such as Ethereum and Hyperledger Fabric, it is clear that when it comes to imposing security regulations, smart contracts are comparable. Programming smart contracts necessitates a distinct technical perspective. Because smart contracts frequently deal with crypto currency, enterprise data and other sensitive information, the consequences of a security breach would be catastrophic. Therefore, special care should be taken into the programming smart contracts, which ensures that traditional security practices are also being followed in smart contract coding. Enforcing rigorous testing of smart contracts before production deployment can minimize the risk of being breached to a certain extent. Further developers should anticipate the security environment around blockchain platforms to shift frequently as new vulnerabilities and security threats are identified and new best practices are developed. Apart from smart contract vulnerabilities discussed in Sections 1.4.1–1.4.3, there are numerous documented weaknesses in Ethereum and Hyperledger Fabric platforms. The scope of this chapter does not allow for consideration of every vulnerability. However, for ease of reference, Table 4.1 shows some of Ethereum vulnerabilities and real-world attacks caused by them. And Table 4.2 illustrates various security risks that could happen in Fabric, its rationale and source.

4.6 ACCESS LAYER ATTACKS IN BLOCKCHAIN

Web2 refers to the current generation of the internet that everyone is familiar with. It shifted the world away from static HTML-based web pages and toward immersive experiences and viewer-generated content. Web3 is a sort of internet that relies on decentralized networks like Ethereum and Bitcoin to function. Web3 is primarily concerned with improving and resolving concerns with current centralized platform-mediated interactions. In Web3 networks, due to its decentralized nature, data resides outside of a centralized server where there is only one point of entry. Beyond the availability of data, there are many security threats in Web3, such as smart contract vulnerabilities, Web3-specific phishing attacks, lack of legal protection if things went wrong, etc. Furthermore, due to Web3's lack of centralized management and data access, policing cybercrime, such as online harassment and extortion, can be considerably more challenging. All these factors, coupled with huge financial incentives linked to Web3 breaches, are a major motivator for attackers to target Web3 applications.

Below sections cover some of the Web3 and Web2 attack vectors adversaries leverage to trick legit users in the blockchain network. Attack vectors discussed here include tactics such as designing malicious dapp, which utilizes unlimited allowances vulnerability, exploitation of third-party APIs to phish users in Web3 application.

Table 4.1 Ethereum vulnerabilities and real-world attacks caused by them

Risk	Rationale	Real-world attack	Source
Reentrancy	Reentrancy occurs when an external callee contract calls back to a function in the caller contract before the caller contract's execution is complete and the internal state of the contract is not updated before passing control to the external contract. Until the caller contract is drained of ether or the transaction runs out of gas, the attacker can get around the due validity check.	The DAO attack	Samreen and Alalfi (2020), Chinen et al. (2020)
Integer overflow	It occurs when the result of an arithmetic operation exceeds the range of a solidity data type, resulting in illegal manipulation of state variables, for example. The flaw is created by solidity code that does not validate numeric inputs properly, and neither the solidity compiler nor the EVM enforces integer overflow/underflow detection.	BEC token attack	Gao et al. (2019)
Delegate call injection	A delegatecall is a particular version of a message call that is similar to a message call except that the code at the destination address is executed in the context of the calling contract and the values of msg.sender and msg.value are not modified. This enables a smart contract to load code from a different address dynamically at runtime. The calling contract is still referenced by storage, current address, and balance.	Parity multi-signature wallet attack (second hack)	Sayeed et al. (2020) SWC-112 · Overview (2022)
Timestamp dependence	It allows the Ethereum network to be decoupled from the globally synchronized clock. For example, to determine the lottery result, a smart contract uses the current timestamp to generate random numbers. Because the smart contract allows miners to post a timestamp within 30 seconds of block confirmation, exploitation opportunities are increased. As a result, the random number generator's output can be changed to obtain advantages.	GovernMental attack	Sayeed et al. (2020) Bartoletti et al. (2017)

(Continued)

Table 4.1 (Continued) Ethereum vulnerabilities and real-world attacks caused by them

Risk	Rationale	Real-world attack	Source
Dependence on transaction order	In Ethereum, miners examine transactions and choose which ones to include in a block based on who has paid a high enough gas price to be included. Furthermore, transactions received to the Ethereum network are transmitted to each node for processing. As a result, an Ethereum node operator may predict which transactions will take place before they are completed. When programming is dependent on the order in which transactions are submitted to it, a race condition vulnerability may occur.	GovernMental attack	Sayeed et al. (2020), SWC-112 · Overview (2022), Bartoletti et al. (2017)
Unlimited nodes creation	A flaw in the node generation mechanism allows an adversary to construct an unlimited number of nodes on a single machine with the same IP address. An attacker with a large number of nodes can take control of the incoming and outgoing connections of victim user nodes, separating them from the rest of the network. The unlimited node creation vulnerability affects Geth clients before version 1.8.	Eclipse attacks	Marcus et al. (2018)
Uncapped incoming connections	Nodes had no upper restriction on the number of inbound TCP connections established by other nodes in Geth clients prior to version 1.8. By initiating a large number of incoming connections to a victim node, an attacker can obscure them. In Geth v1.8, the vulnerability is mitigated by imposing a restriction on the amount of incoming TCP connections to a node.	Eclipse attacks	Marcus et al. (2018)

4.6.1 Exploitation via Web3 attacks

While bugs and vulnerabilities can happen even in reputable and legitimate decentralized applications (dapps), there have also been cases where the dapp itself was malicious. Designing a user interface of the infamous Ethereum-based "yield farming platform" scam known as UniCats (Be'ery, 2020) with frontend features similar to other popular De-Fi projects at that time allowed hackers behind UniCats to manipulate victim users into believing that UniCats is legitimate and trustworthy when in truth hackers

Table 4.2 Security risks in Hyperledger Fabric

Risk	Rationale	Source
Using getTxTimestamp() value as a seed to PRNG function	Malicious user who has access to both transaction timestamp and the smart contract will be able to recompute the random number and thus compromise the security of smart contract logic.	Paulsen et al. (2021)
Use of map structure iteration	Non-determinism may result from the use of map structure iteration. Developers who are familiar with the Go standard or implementation can avoid this issue.	Yamashita et al. (2019), Brotsis et al. (2020)
Manipulation of variable values via a pointer	A pointer is a memory address that fluctuates depending on the environment. As a result, adopting reified object addresses may result in non-determinism.	Yamashita et al. (2019), Brotsis et al. (2020)
State-changing transactions depends on a global variable	Attacker forces smart contract to be non-deterministic or behave unexpectedly.	Paulsen et al. (2021)
Concurrency	Concurrency is well supported in Go, thanks to goroutine and channel. If several transactions are processed at the same time and under heavy loads, a change in the versions of the keys could result in key collisions and double spending.	Yamashita et al. (2019), Brotsis et al. (2020)
Range query risk	If an asset retrieved from a StateDatabase, such as CouchDB, is not checked against the most recently committed ledger value, an attacker could cascade unauthorized modifications across the network.	Paulsen et al. (2021)
Usage of third-party APIs	When third-party APIs are used and if they return different results to each peer, the ledgers become inconsistent.	Yamashita et al. (2019)

was abusing transferFrom (sender, recipient, amount) function of ERC20 standard for exploiting users. transferFrom() function moves the specified amount of tokens from sender to recipient using the allowance mechanism; transferred amount is then deducted from the caller's allowance (ERC 20- OpenZeppelin Docs 2022). In UniCats, hackers implemented transferFrom() function such that it would request an unlimited allowance from the victim user. While this dapp behavior offered a superior user experience (since the user does not need to approve a new allowance every time when they want to deposit tokens), in reality, when a user allows unlimited allowance to UniCats, it was gaining permission to spend tokens on its user's behalf. When the malicious UniCats was rug-pulled, the scammers not only took the deposited funds but also all tokens that victim users had in their wallets.

4.6.2 Exploitation via Web2 attacks

It is not necessary for an attacker who wishes to target Ethereum blockchain users to use or exploit Ethereum/Web3 (Nath et al., 2020) specific security issues. Due to the lack of laws enforced in public blockchains such as Ethereum, an attacker can instead employ his or her expertise in typical Web2 attack vectors such as phishing, Ponzi schemes and so on. Since 2017, phishing scams have produced more than half of all cybercrime revenue on Ethereum, according to a report by "Chainalysis" (The Rise of Cybercrime on Ethereum – Chainalysis 2022). In traditional Web2 phishing attacks, an adversary sets up a duplicate of the legitimate website and entices victims to perform actions such as checking in, etc., on the phony attacker-created website in order to collect sensitive information such as user passwords, payment information and so on. The hack happened on Badger DAO on 2nd December 2021, where hackers managed to steal up to $120 million of user's funds (BadgerDAO Exploit Technical Post Mortem 2022), is a well-suited example to demonstrate how adversaries could leverage Web2 vectors to compromise Web3 application.

Badger DAO is a decentralized autonomous organization that enables users to use their Bitcoin as a collateral across De-Fi applications. In Badger DAO hack, hackers never exploited security flaws in smart contracts; rather they relayed on phishing and other Web2 attack vectors such as improper access control for exploitation. The attack happened in the below sequence of events.

- Initially, attackers took leverage of vulnerability in Cloudflare Workers (BadgerDAO Exploit Technical Post Mortem 2022) (an application platform that runs on Badger's cloud network), which would allow attackers to create and view (Global) API keys before email verification was completed.
- Attackers used this compromised API key from Cloudflare Workers for injecting malicious scripts into the html of app.badger.com.
- The script intercepted web3 transactions and prompted users to allow a foreign address approval to operate on ERC-20 tokens in their wallet.
- In order to avoid detection, attackers deployed and removed the malicious script periodically. They only targeted wallets over a certain balance and accessed the API from multiple proxies and VPN IP addresses

The hackers took $130 million in total; however, about $9 million was recovered since the assets had been transferred but not yet withdrawn from Badger's vaults.

4.6.3 Best practices

Cybersecurity is a never-ending process of development and an arms race between attackers and defenders. In a Web3-based internet, some of the traditional best practices that have accumulated through time still stand up and therefore should be maintained. And some tried-and-true strategies, such as encryption/decryption and zero-trust frameworks, can be adapted and used effectively. Most often organizations may adopt a blockchain solution which is integrated with other Web2 components like cloud, enterprise on-prem solutions, etc. In such scenarios, organizations also need to look into securing their solution from Web2 attack vectors such as phishing, API vulnerabilities, network layer attacks, etc.

4.7 APPLICATION LAYER ATTACKS IN BLOCKCHAIN

As the saying goes "A Chain is as strong as the weakest link", organizations cannot only relay on security measures offered/implemented in blockchain; rather, they need to adopt a holistic approach for the employment of security across the solutions. Most often, it's the end-user application which is considered to be the weakest in terms of security in a blockchain solution. The main reason for this that the end-user application serves as the primary channel to access the solution. Even with blockchain platforms offering their consensus protocols and cryptography to ensure the integrity of ledger, it is still possible for an adversary to exploit a blockchain-based application with minimum effort if the end-user application is designed with poor security controls. Organizations who design blockchain solutions should be aware of how adversaries trigger well-known attack vectors from the end-user interface for compromising blockchain solutions.

Below sections cover some of the user interface attack vector adversaries leverage to exploit victim users of a blockchain application. Attack vectors (cross-site scripting, weak password complexity and unvalidated URL redirection) discussed here include tactics such exploiting the code/design flow of the frontend user interface for gaining sensitive information or access to the blockchain-based application.

4.7.1 Cross-site scripting (XSS)

Cross-site scripting (XSS) attacks are injection attacks in which malicious scripts are inserted into otherwise trustworthy and harmless websites. XSS attacks occur when an attacker utilizes a web application to transmit malicious code to a separate end user, usually in the form of a browser-side script. The browser of the end user has no means of knowing that the malicious script from attackers should not be trusted and will run it, nonetheless. The malicious script can access user cookies, tokens that maintain user sessions or other sensitive information stored by the browser since it believes that the script came from a trusted source. These malicious scripts injected by attackers can even rewrite the HTML page's structure/content (Singh et al., 2020).

In Ethereum realm, XSS was first observed against EtherDelta crypto-currency exchange (How one hacker stole thousands of dollars' worth of cryptocurrency with a classic code injection|HackerNoon 2022). For signing the token trade transaction, EtherDelta requires users to import their private key and account address into their web browser. EtherDelta also posted the newly formed token's name on its website, which was derived from the token contract's code. The attacker took advantage of EtherDelta's token name rendering functionality by uploading a new token contract with malicious JavaScript code in the token's name, which when rendered on the user's browser got executed and stole user's private key from the browser (Chen et al., 2020). The use of suitable validation and encoding of user inputs, as well as the usage of appropriate response headers, can help to prevent XSS attacks (Cross-Site Scripting Prevention – OWASP Cheat Sheet Series 2022).

4.7.2 Weak password complexity

To have an assertion of authenticity for a particular user in a system, authentication systems frequently rely on a stored secret/password. As a result, it's critical that this password be sufficiently complicated and difficult to guess for an attacker. A weak password is one that is short, common, a system default or one that can be quickly determined using a subset of all possible credentials, such as dictionary terms, full nouns, keywords based on the username or popular variations on these subjects (Dell'Amico et al., 2010). Examples of weak passwords include "123456", "password", "admin" and "qwerty". An attacker could compromise a vulnerable application, which does not enforce strong password complexity via variety of attacks which include brute forcing, social engineering, etc.

Enigma, a decentralized investing platform, was hacked ahead to its initial coin offering (ICO) in August 2017 because of a weak password vulnerability (Jae Hyung, 2019). The adversary employed social engineering to steal an Enigma founder's passcode, which had previously been revealed in another breach and had been repeated in Enigma. As a result, the hacker gained access to the company's Slack channel, mailing lists and the Google account that hosted the ICO presale. Then, the attacker changed the legitimate ICO contract address to his or her own address and sent messages to solicit buyers in phoney presales. The authentication mechanism's whole success depends on choosing the right password policies and imposing them via implementation. The common ways to reduce the risk of readily guessed passwords aiding unlawful access are to introduce new authentication restrictions (such as two-factor authentication) or implement a strong password policy (Web Security Testing Guide (WSTG) – Latest|OWASP Foundation 2022).

4.7.3 Unvalidated URL redirection

Unvalidated URL redirection is a vulnerability that allows the attacker to redirect victim users to an untrusted external site. The most common method of attack is to provide a link to the victim, who subsequently clicks it and is unwittingly forwarded to the malicious attacker-hosted website. Thus, unvalidated URL redirection vulnerability can be used to make phishing attacks against users easier. Because most users may not notice the following redirection to a different but malicious domain, the ability to utilize an authentic application URL, targeting the right domain, and with a valid SSL certificate adds legitimacy to the phishing attempt. In phishing attacks involving unvalidated URL redirection vulnerability, once on the attacker's phishing site, users enter their credentials on the login form presented to them, which directs them to a script managed by the attacker. The script is often used to steal the victim user's login credentials as they were entered in and saved on the server. Attackers subsequently utilize these stolen credentials to impersonate the victim users on the actual website (Chen & Freire, 2021).

A combination of unvalidated URL redirection, Border Gateway Protocol (BGP) and DNS hijacking attack was observed in MyEtherWallet on April 2018 (Poinsignon, 2018), which resulted in an overall loss of US$17 million from the victim user's wallets. Applications should avoid including user-controllable data in redirection targets. If receiving user-controllable input is essential for the redirection feature, sanitize the user input and ensure that the provided value is acceptable for the application and allowed for the user (Unvalidated Redirects and Forwards – OWASP Cheat Sheet Series 2022).

4.7.4 Best practices

The attack vectors covered in Sections 1.6.1–1.6.3 are those that are straightforward to fix/avoid from a mitigation standpoint while also being easy for the attacker to exploit if they exist. Even a novice hacker could carry out XSS, URL redirection and other attacks. The ease with which these vulnerabilities can be exploited, as well as the financial and reputational impact they have on businesses, make them extremely risky. To address these flaws, businesses must ensure that established security practices are used in their solutions, as well as performing thorough application testing to uncover security flaws. OWASP's Application Security Verification Standard (ASVS) (OWASP Application Security Verification Standard|OWASP Foundation 2022) and WSTG (WSTG – v4.1|OWASP Foundation 2022) are two of the most widely used resources in the security industry, providing developers and security professionals with a list of requirements for secure development, technical security controls, guidance on detecting security flaws in applications, etc.

4.8 SECURITY STRATEGY FOR BLOCKCHAIN DEVELOPMENT

As outlined in the previous sections, blockchain solutions should incorporate security controls across different layers to provide protection from various security threats. Software development community has realized that following SSDLC practices is the optimal approach to increase the resiliency of software systems while reducing costs associated with fixing security bugs. SSDLC emphasizes on baking in security practices across various stages of development instead of bolting on security layer or activities on a finished product. This is relevant to blockchain applications/solutions as well.

Typical development lifecycle for blockchain solution comprises of several phases:

1. **Plan & Define:** During the plan and define phase, business requirements are understood and captured in different representations such as user stories, use cases, etc.
2. **Design:** Design phase involves the creation of architectural, deployment diagrams and detailed design documentation.
3. **Develop & Build:** Design documents and wireframes are translated to actual code based on the blockchain platform.
4. **Test:** Developed solution is tested to uncover functional errors, performance bottlenecks and to confirm compliance with business requirements.
5. **Deployment and Maintenance:** Once the solution is deployed, the solution needs operational and maintenance support.

Security practices should be baked across the development cycle to create resilient blockchain solutions while minimizing the security cost. During the planning phase, a security architect can help to capture security and compliance requirements along functional business requirements. Once the design of the solution is drafted, the design can be evaluated for security flaws on the architectural level. Architecture security review and threat modeling are valuable in understanding and emulating high-level threats and design flaws before even the first line of code is written.

Once the smart contract and associated application code are written, the source code can be analyzed for security vulnerabilities. While there is extensive tool support for conventional applications and API code, there are limited tools with support for smart contract review. Furthermore, smart contract vulnerabilities are often deeply connected with the business use cases and require manual review by an expert. Organizations should consider performing penetration testing on the blockchain solution when it is deployed in a test environment. Penetration testing can add additional value in uncovering runtime vulnerabilities and help to perform impact analysis on vulnerabilities discovered during code review.

Finally, once the system is deployed to end users, security should be factored into the ongoing maintenance and support. Architects can use various security technologies such as network firewalls, web application firewalls (WAF) and runtime application self-protection (RASP) solutions to ensure that the deployed solution is resilient to attacks.

4.9 CONCLUSION

The adoption of blockchain has its challenges, and organizations must secure and protect their blockchain infrastructure. As blockchain technology evolves, so are adversary tactics for compromising blockchain-based systems. The goal of this chapter was to demonstrate the diversity of adversary tactics used against blockchain solutions at each layer, including the network layer, smart contracts, access layer, application layer and establish a strategy for designing, developing and deploying secure blockchain solutions. As blockchain technology continues to disrupt global industries, organizations must be cognizant of the security implications and attack surface of blockchain. Organizations with a security-infused blockchain strategy are more likely to realize long-term return of investment (ROI) while minimizing cyber risks.

REFERENCES

Anita, N. and Vijayalakshmi, M. (2019). Blockchain security attack: A brief survey. *2019 10th International Conference on Computing, Communication and Networking Technologies (ICCCNT).* https://doi.org/10.1109/ICCCNT45670.2019.8944615.

BadgerDAO Exploit Technical Post Mortem. (n.d.). https://badger.com/technical-post-mortem.

Bartoletti, Massimo., Carta, Salvatore., Cimoli, Tiziana., and Saia, Roberto. (2017). Dissecting ponzi schemes on ethereum: Identification, analysis, and impact. *Future Generation Computer Systems*, 102, 259–277. https://doi.org/10.1016/j.future.2019.08.014.

Be'ery, T. (2020). *UniCats Go Phishing.* zengo. https://zengo.com/unicats-go-phishing/.

Brotsis, Sotirios, Kolokotronis, Nicholas, Limniotis, Konstantinos, Bendiab, Gueltoum, and Shiaeles, Stavros. (2020). *On the Security and Privacy of Hyperledger Fabric: Challenges and Open Issues. 2020 IEEE World Congress on Services (SERVICES).* https://doi.org/10.1109/SERVICES 48979.2020.00049.

Channels – Hyperledger-fabricdocs Main Documentation. (n.d.). https:// hyperledger-fabric.readthedocs.io/en/latest/channels.html.

Chen, Huashan, Pendleton, Marcus, Njilla, Laurent, and Xu, Shouhuai. (2020). A survey on ethereum systems security: Vulnerabilities, attacks and defenses. *ACM Computing Surveys,* Vol. 53, No. 3, Article 67. https://doi. org/10.1145/3391195.

Chen, Zhouhan, and Freire, Juliana. (2021). Discovering and measuring malicious URL redirection campaigns from fake news domains. *2021 IEEE Security and Privacy Workshops (SPW).* https://doi.org/10.1109/SPW53761.2021.00008.

Chinen, Yuichiro, Yanai, Naoto, Paul Cruz, Jason, & Okamura, Shingo. (2020). Hunting for re-entrancy attacks in ethereum smart contracts via static analysis. *IEEE Blockchain 2020.* https://arxiv.org/abs/2007.01029.

Cross Site Scripting Prevention – OWASP Cheat Sheet Series. (n.d.). https://cheat-sheetseries.owasp.org/cheatsheets/Cross_Site_Scripting_Prevention_Cheat_ Sheet.html.

Dabholkar, Ahaan, and Saraswat, Vishal. (2019). Ripping the fabric: Attacks and mitigations on hyperledger fabric. *Applications and Techniques in Information Security, 10th International Conference, ATIS 2019, Thanjavur, India, November 22–24, 2019.* http://dx.doi.org/10.1007/978-981-15-0871-4_24.

Davenport, Amanda, Shetty, Sachin, and Liang, Xueping. (2018). Attack surface analysis of permissioned blockchain platforms for smart cities. *2018 IEEE International Smart Cities Conference (ISC2).* https://doi.org/10.1109/ ISC2.2018.8656983.

Dell'Amico, Matteo, Michiardi, Pietro, and Roudier, Yves. (2010). *Password Strength: An Empirical Analysis. 2010 Proceedings IEEE INFOCOM.* https://doi.org/10.1109/INFCOM.2010.5461951.

Ehab, Zaghloul, Tongtong, Li, Matt, W. Mutka, and Jian, Ren. (2020). Bitcoin and blockchain: Security and privacy. *IEEE Internet of Things Journal.* https:// doi.org/10.1109/JIOT.2020.3004273.

ERC 20- OpenZeppelin Docs. (n.d.). https://docs.openzeppelin.com/contracts/2.x/ api/token/erc20#IERC20-transferFrom-address-address-uint256-.

Eskandari, Shayan, Moosavi, Seyedehmahsa, and Clark, Jeremy (2020). *SoK: Transparent Dishonesty: Front-Running Attacks on Blockchain.* Springer, Cham. https://doi.org/10.1007/978-3-030-43725-1_13.

Fabric CA User's Guide - Hyperledger-Fabric-Cadocs Main Documentation. (n.d.). https://hyperledger-fabric-ca.readthedocs.io/en/latest/users-guide.html# fabric-ca-server.

Fabric-Samples/Compose-Couch.Yaml. (n.d.). *At Main·Hyperledger/fabric-samples.* GitHub. https://github.com/hyperledger/fabric-samples/blob/main/ test-network/compose/compose-couch.yaml.

Forecast: Blockchain Business Value, Worldwide, 2017–2030. (n.d.). Gartner. https://www.gartner.com/en/documents/3627117/forecast-blockchain-business-value-worldwide-2017-2030.

Gao, Jianbo, Liu, Han, Liu, Chao, Li, Qingshan, Guan, Zhi, & Chen, Zhong (2019). EASYFLOW: Keep ethereum away from overflow. *2019 IEEE/ACM 41st International Conference on Software Engineering: Companion Proceedings (ICSE-Companion).* https://ieeexplore.ieee.org/document/8802775.

How One Hacker Stole Thousands of Dollars Worth of Cryptocurrency with a Classic Code Injection. HackerNoon. (n.d.). https://hackernoon.com/how-one-hacker-stole-thousands-of-dollars-worth-of-cryptocurrency-with-a-classic-code-injection-a3aba5d2bff0.

Introduction to Smart Contracts. Ethereum.org. (2022). https://ethereum.org/en/developers/docs/smart-contracts/.

Iqbal, Mubashar, and Matulevicius, Raimundas. (2021). Exploring Sybil and double-spending risks in blockchain systems. *IEEE Access*, 9. https://doi.org/10.1109/ACCESS.2021.3081998.

Jae Hyung, Lee. (2019). *Systematic Approach to Analyzing Security and Vulnerabilities of Blockchain Systems.* Massachusetts Institute of Technology. https://hdl.handle.net/1721.1/121793.

Yongrae, Jo, and Chanik, Park. (2019). BlockLot: Blockchain based verifiable lottery. *arXiv preprint arXiv:1912.00642.*

Marcus, Y., Heilman, E., and Goldberg, S. (2018). Low-resource eclipse attacks on Ethereum's peer-to-peer network. *IACR Cryptology ePrint Archive*, 2018.

Mastilak, Lukas, Galinski, Marek, Helebrandt, Pavol, and Kotuliak, Ivan. (2020). Enhancing border gateway protocol security using public blockchain. *Sensors*, 20(16): 4482. http://dx.doi.org/10.3390/s20164482.

Nath, Keshab, Dhar, Sourish, and Basishtha, Subhash. (2014). Web 1.0 to Web 3.0 – Evolution of the web and its various challenges. In *2014 International Conference on Reliability Optimization and Information Technology (ICROIT).*

Networks. Ethereum.org. (2022). https://ethereum.org/en/developers/docs/networks/.

OWASP Application Security Verification Standard. OWASP Foundation. (n.d.). https://owasp.org/www-project-application-security-verification-standard/.

Paulsen, Cathrine, Liang, Kaitai, and Chen, Huanhuan (2021). Revisiting smart contract vulnerabilities in hyperledger fabric. *TU Delft Electrical Engineering, Mathematics and Computer Science*; TU Delft Intelligent Systems. http://resolver.tudelft.nl/uuid:dd09d153-a9df-4c1b-a317-d93c1231ee28.

Poinsignon, Louis. (2018). *BGP Leaks and Cryptocurrencies.* cloudflare. https://blog.cloudflare.com/bgp-leaks-and-crypto-currencies/.

Private Data – Hyperledger-Fabricdocs Main Documentation. (n.d.). https://hyperledger-fabric.readthedocs.io/en/latest/private-data/private-data.html.

Samreen, Noama Fatima, and Alalfi, Manar H. (2020). Reentrancy vulnerability identification in ethereum smart contracts. *2020 IEEE International Workshop on Blockchain Oriented Software Engineering (IWBOSE).* https://doi.org/10.1109/IWBOSE50093.2020.9050260.

Sayeed, Sarwar, and Marco-Gisbert, Hector. (2019). Assessing blockchain consensus and security mechanisms against the 51% attack. *Applied Sciences.* https://doi.org/10.3390/app9091788.

Sayeed, Sarwar, Marco-Gisbert, Hector. and Caira, Tom. (2020). Smart contract: Attacks and protections. *IEEE Access.* https://doi.org/10.1109/ACCESS.2020.2970495.

Singh, Mehul, Singh, Prabhishek, and Kumar, Pramod. (2020). An analytical study on cross-site scripting. *2020 International Conference on Computer Science, Engineering and Applications (ICCSEA)*. https://doi.org/10.1109/ICCSEA49143.2020.9132894.

Smart Contracts and Chaincode - Hyperledger-fabricdocs Main Documentation. (n.d.).   https://hyperledger-fabric.readthedocs.io/en/latest/smartcontract/smartcontract.html.

Sukhwani, Harish, Wang, Nan, Trivedi, Kishor S, and Rindos, Andy. (2018). Performance modeling of hyperledger fabric (permissioned blockchain network). In *IEEE 17th International Symposium on Network Computing and Applications (NCA)*.

SWC-101 · Overview. (n.d.). https://swcregistry.io/docs/SWC-101#remediation.

SWC-107 · Overview. (n.d.). https://swcregistry.io/docs/SWC-107#remediation.

SWC-112 · Overview. (n.d.). https://swcregistry.io/docs/SWC-112.

SWC-114 · Overview. (n.d.). https://swcregistry.io/docs/SWC-114.

The Operations Service. (n.d.). *Hyperledger-Fabric*. Retrieved January 15, 2022, from https://hyperledger-fabric.readthedocs.io/en/latest/operations_service.html.

The Rise of Cybercrime on Ethereum – Chainalysis. (2017). https://blog.chainalysis.com/reports/the-rise-of-cybercrime-on-ethereum/.

Transaction Flow – Hyperledger-Fabricdocs Main Documentation. (n.d.). https://hyperledger-fabric.readthedocs.io/en/latest/txflow.html.

UniCat – Earn MEOW By Staking UNI. (n.d.). https://web.archive.org/web/20200921225451/.

Unvalidated Redirects and Forwards – OWASP Cheat Sheet Series. (n.d.). https://cheatsheetseries.owasp.org/cheatsheets/Unvalidated_Redirects_and_Forwards_Cheat_Sheet.html.

WSTG – Latest | OWASP Foundation. (n.d.). *WSTG – Latest*. OWASP Foundation. https://owasp.org/www-project-web-security-testing-guide/latest/4-Web_Application_Security_Testing/04-Authentication_Testing/07-Testing_for_Weak_Password_Policy.

WSTG – V4.1 | OWASP Foundation. (n.d.). *WSTG – V4.1*. OWASP Foundation. https://owasp.org/www-project-web-security-testing-guide/v41/2-Introduction/.

Yamashita, Kazuhiro, Nomura, Yoshihide, Zhou, Ence, Pi, Bingfeng, and Jun, Sun. (2019). Potential risks of hyperledger fabric smart contracts. *2019 IEEE International Workshop on Blockchain Oriented Software Engineering (IWBOSE)*. https://doi.org/10.1109/IWBOSE.2019.8666486.

Yang, Xinle, Chen, Yang, and Chen, Xiaohu. (2019). Effective scheme against 51% attack on proof-of-work blockchain with history weighted information. *2019 IEEE International Conference on Blockchain (Blockchain)*. https://doi.org/10.1109/Blockchain.2019.00041.

## Chapter 5

Problems with the implementation of blockchain technology for decentralized IoT authentication

A literature review

Vijay Anant Athavale
Walchand Institute of Technology

Ankit Bansal
Chitkara University

CONTENTS

DOI: 10.1201/9781003282914-5

5.1 INTRODUCTION

As the internet expands, brand new technologies and adaptations of old technologies happen over time. The blockchain technology was created to allow secure and privacy-focused communication over a public medium where stored data is shared and completely transparent to every user. This is quite a contradiction, which is why people have been trying to figure out what other uses there could be for such technology rather than just handling currency. The Internet of Things (IoT) can be seen as an expansion of the internet that allows things to access it that they otherwise would not be able to. With the addition of internet access, some security problems as well as questions about how to scale the networks as more and more things become connected arise.

One of the proposed solutions to these IoT problems is the amalgamation of blockchain and IoT. A secure and privacy-focused decentralized solution is what is badly needed in the IoT sphere. But there is a big contradiction between these technologies. Blockchain technology is known for being an extremely resource-heavy technology where a lot of computational power and storage are needed to operate it to its full potential. This is in stark contrast to the IoT, which is mainly a bunch of low-powered devices that need to keep energy usage to an absolute minimum to be able to be deployed in a manner where they are useful.

The early solutions using the blockchain also correctly identified latency as a big problem, as blocks in that chain are only written about every 10 minutes. Some progress has been made in this regard by the cryptocurrency blockchains themselves, as newer ones like Ethereum write a block about every 12 seconds, which is a huge improvement. However, it is still not enough to become a competitive IoT solution, let alone a standard. Profit-driven entities would have some trouble selling the "new revolutionary solution that processes your actions after 12 seconds".

It is apparent that there are some big road bumps before these technologies can work in harmony, and no defacto solution to either of these problems has seemingly been identified. This is where the present study comes in to hopefully help steer the research efforts in the right direction.

This study aims to expose the most common problems within proposed solutions for merging these technologies. It will be done in the form of a systematic literature review where solutions are identified, categorized, and then criticized to see if there are any patterns in what researchers believe will work or what problems researchers collectively fail to fix. The idea is to guide future research into either what types of solutions seem collectively most probable or what is keeping a specific group of solutions from reaching their full potential.

The rest of the chapter is organized as follows. In Section 5.2, we discuss the background of our research. This section consists of two major parts.

We detail the emergence of the IoT (Section 5.2.1) and the emergence of blockchain technology (Section 5.2.2). In Section 5.3, we discuss various consensus algorithms in detail. In Section 5.4, we list some public key infrastructures (PKIs). In Section 5.5, we list some issues related to the integration of blockchain technologies with information technology (IT) infrastructures. In Section 5.6, we detail our problem definition and motivation. In Section 5.7, we discuss the methodology to answer the problem definition. In Sections 5.8–5.10, we discuss various search terms, data extraction and analysis, and searches and paper assessment. In Section 5.11, we discuss cluster-based solutions for various applications. In Sections 5.12–5.16, we discuss the use and issues faced in changing PKI, the use of blockchain servers, adaptation of blockchain technology, device-to-device communication, and key management in the case of third-party authentication. In Sections 5.17 and 5.18, we summarize and discuss the identification of problems. Finally, we conclude the chapter in Section 5.19 and discuss future work in Section 5.20.

5.2 BACKGROUND

This section describes what IoT and blockchain technology are. It also explains some concepts of a fully functioning IoT system which can be tied to the benefits of blockchain technology and gives an understanding of why these technologies can work together.

5.2.1 Internet of things

IoT is a constantly growing technology. While it is growing, new challenges that have to be solved appear, and old solutions can become obsolete because of the new requirements. IoT is basically the act of adding internet capabilities to things that traditionally would not have them. Essentially, all "smart" items like smart watches, smart fridges, smart TVs, etc. Jayavadhana et al. (2013) have brought forward four definitions of what IoT is:

> The worldwide network of interconnected objects is uniquely addressable based on standard communication protocols.

Things are active participants in business, information, and social processes where they are enabled to interact and communicate among themselves and with the environment by exchanging data and information sensed about the environment while reacting autonomously to real-world events and influencing them by running processes that trigger actions and create services with or without direct human intervention.

IoT uses information and communications technologies to make the critical infrastructure components and services of a city's administration, education, healthcare, public safety, real estate, transportation, and utilities more aware, interactive, and efficient.

Interconnection of sensing and actuating devices provides the ability to share information across platforms through a unified framework, developing a common operating picture for enabling innovative applications. This is achieved by seamless ubiquitous sensing, data analytics, and information representation with cloud computing as the unifying framework.

For an IoT system to be fully operational, it will require three different components. First is hardware, which is the sensors, actuators, embedded communication hardware, etc. Second is middleware, which consists of on-demand storage for all the data gathered from the hardware and tools for analytics of said data. Third and last is presentation, which consists of human-readable interpretations of the stored and refined data and can be accessed on different platforms and applications.

An example of hardware is radio frequency identification (RFID). A passive RFID tag has the benefit of not needing electricity in the object itself, but instead works like a wireless barcode and uses the power emitted from the reader to communicate the ID of the object. There is also the possibility of active tags with a power source that can start authentication themselves rather than just respond. One current implementation of passive tags is bank cards, which can be used wirelessly (Jayavadhana et al., 2013).

Middleware consists of low-powered sensors that collect data from the edge hardware, among other things, in what is called a wireless sensor network (WSN). Sensors, just like edge devices, will most likely have restrictions, which means that they cannot store all the data themselves. Therefore, sensor data is sent to a distributed or centralized system for analytics. The middleware itself consists of the following four parts:

Hardware consists of interfaces for edge sensors, processing units for handling and potentially calculating the collected data, transceiver units to be able to send the processed data forward and receive data from the sensors, and finally a power supply to give the parts the power they require. But that is not enough for the middleware to fulfil its purpose in the IoT system. It also needs networking between the nodes. For most applications, they should be set up in an ad-hoc manner. Ad-hoc means a decentralized approach where the nodes all handle routing together by communicating with each other instead of the traditional use of routers or access points. The point of the routing is to find the fastest way through the network to a base station called a sink node that acts as a layer between the WSN and the internet to allow access to the data from anywhere. Third, the middleware itself needs middleware that translates the data collected to a state that is readable for other programmes and, by that, becomes readable by humans and manageable. Lastly, secure data aggregation is needed as the middleware needs a way to ensure reliable data without putting too much strain on the network as node failures are common in WSNs. Security is also a huge point here, as it is to ensure that only authorized sensors and nodes can handle and create data within the network (Jayavadhana et al., 2013).

Finally, the system needs an addressing scheme to help identify every single "thing" in the system, making it possible to know which sensor or node is sending what data or to potentially control something from the internet. IPV4 can work to some extent to help identify at least a group of devices and their location. However, it is not enough to individually identify devices because of the number of devices in a network. To help with this, a system like uniform resource name (URN) can be used. The idea is then to use the URN system for the internal subnets behind the sink node while using regular IP for the outside. This would mean that a sink node has to be assigned to a specific subnet and all devices connected to it have to be added to a lookup table to forward the incoming data to the right device. This all together then forms an IoT system that can be accessed from the internet and has the possibility to address every single device within it (Jayavadhana et al., 2013).

5.2.2 Blockchain technology

This technology is pretty self-explanatorily and, just like its name suggests, is a database created by a sequence of data blocks. The reason it is considered a chain is because all new data blocks added to the database have to refer to the last added block already in the database. This means that the data in the chain is in chronological order. All data in a single block is considered to be added at the exact same time (Crosby et al., 2016).

Hongning et al. (2017) define four key characteristics of blockchain technology:

Unlike a centralized approach, there is no need to validate data through a single point, which could act as a bottleneck in the system. Third parties are therefore completely eliminated for transactions between nodes in a blockchain. How the validity of blocks is handled can be found in Section 5.2.1.

Transactions can be validated quickly, and with an incentive behind mining, invalid transactions are less likely. Deleting or changing old transactions is nearly impossible as the entire block and every single one behind it in the chain would have to be revalidated. Invalid transactions can be discovered immediately.

Instead of using a real identity, each user of the blockchain only has to use a randomly generated address. There are, however, other ways to identify a user, both by the nature of how the blockchain nodes are handled and the shortcomings of the regular internet.

As demonstrated by the Bitcoin blockchain, a system like unspent transaction output can be used for transparency. Basically, the blockchain contains already made transactions that have an unspent status, and when an actual new transaction is made, it refers to one of these, which changes its status to spend and closes it for further transactions. A new unspent transaction is then created instead, which further new transactions can use. This unspent/spent then allows for transactions to be easily verified and tracked.

Pilkington (2016) explains that there are three different privacy levels for blockchains, which essentially are levels of decentralization. A public blockchain is completely decentralized and is available to every internet user. For these blockchains, a consensus has to be reached by a majority of participating computers to allow a new block to be added to the chain. A private blockchain instead has a centralized approach where a chosen group handles and monitors write permissions to the blockchain. The read permissions for the blockchain can also be restricted if the public should not be allowed any access at all. In an organizational approach, this would allow for whitelisting and blacklisting of user identities, something that is not available for a public blockchain. The last type is a hybrid between the two that is partially decentralized and is called a consortium blockchain. Essentially, multiple organizations have access to the blockchain instead of only one, as in the private scenario. Hongning et al. (2017) further explain that public blockchains have the upside of being almost completely tamper-proof at the expense of efficiency in creating the blocks, while private and consortium blockchains instead have higher efficiency but have a higher risk of data being tampered with.

As explained by Hongning et al. (2017), each block in the chain consists of a block header and a block body. The block header itself contains several pieces of information:

Block Version: The set of rules to be followed for new block validation.
A Merkle tree root hash is a hash value derived from all of the data in the block.
A timestamp in UNIX.
nBits: A value that the newly calculated hash for the specific block has to be lower than to be considered a valid solution.
Nonce: A 32-bit number that in some algorithms starts with zero and is increased every time a hash is calculated or has to be determined to create a hash that is below the nBits value and therefore makes the block valid. Which statement is true depends on the consensus algorithm used by the blockchain.
Parent Block Hash: A hash that points to the block that precedes the current block in the chain.

The block body simply contains the actual data in the form of transactions, which each would be one instance of whatever data is entered into the blockchain and a field containing the number of transactions in the block. The maximum number of transactions a block can handle is determined by the block size and the transaction size.

As shown by Cho et al. (2017), the newer blockchains like Ethereum have the functionality to write smart contracts, which basically means developers can write programmes that run on top of Ethereum. By the nature of blockchains, developers can assume that the programme in the smart contract cannot be changed without permission but can still be read from anywhere. This means Ethereum works as a massive shared computing system. This is one of the reasons people have realized that blockchain technology can be used for more than just cryptocurrency and is, for now, a major stepping stone in trying to create new solutions based on blockchains.

5.3 CONSENSUS ALGORITHMS

Hongning et al. (2017) explain that with the decentralized nature of blockchains, with no single node to control that every block added is trustworthy; new solutions had to be developed for all participants to reach a joint conclusion. Six different algorithms that achieve this goal have been identified.

Proof of work (PoW) is the strategy used for consensus in the Bitcoin network. A decentralized network still needs someone to record transactions, and a preferred method would be random selection. This, however, could lead to malicious actors being chosen from the random draws. Instead, the need to solve an artificial workload is put up so that someone that wants to add blocks to the chain has to work their way through. Generally, and in the case of Bitcoin, the workload is calculations. The value that should be calculated is a hash that is created from the collective values in the header of the block. As the nonce is the only value in the header that is not set, this is what would have to be changed and calculations would have to be done for each possible nonce value until a hash that is equal to or below the value of the nBits. Once a node has calculated one of these valid hashes, it will send the block with the correct nonce value to all the other nodes. Collectively, the calculation of the hash is checked, and if the given solution checks out on the other nodes as well, it is added to the blockchain. With the decentralized nature of a blockchain that would utilize this strategy, two correct hashes could be identified almost at the same time. This creates two branches of the blockchain and nodes will keep working on both of these branches. The branch that gets a new validated block first is then chosen as the real blockchain, while the alternate branch is discarded, causing the transactions in the discarded branch to have to be revalidated in the new blockchain. The nature of this solution means most nodes trying to calculate spend most of their resources calculating completely useless values, essentially wasting power (Hongning et al., 2017).

Proof of stake (PoS) tries to cut back on the energy waste of the PoW algorithm. This is a cryptocurrency-targeted algorithm that instead of having nodes prove themselves by doing calculations, lets them prove themselves by having a lot of currency. The reasoning is that a node with a lot invested in the blockchain would not want to harm it and thus the currency's value. This will end up in an unfair "rich get richer" situation as the node with the most currency also has the highest chance of creating more of it. To combat this, solutions that combine the currency with something else to determine the node that gets to create the next block have been created. Blackcoin uses a randomized approach where a formula looks at a hash value that is created for each one of the nodes together with the amount of currency the node has to determine who gets to add a block. Peercoin instead values the age of the currency as well as its size to give a node a better chance of getting the correct calculation. While these solutions decrease the energy consumption and increase the efficiency of the blockchain compared to the PoW approach, attacks might come as a consequence since the power to calculate a new block is easier to come by (Hongning et al., 2017).

Practical byzantine fault tolerance is an algorithm that allows for one-third of the nodes to not be in agreement with the rest but still send through a block. Blocks are created in what is called a "round," where a primary node is selected based on some set rules. This primary node is then in control of creating the block. To create a block, three phases have to be completed, where two-thirds of the nodes have to vote it through each time. Since this solution is based on a specific number of nodes, all nodes must be known in the network for it to work (Hongning et al., 2017).

Delegated PoS is a version of PoS where the nodes holding currency instead choose representatives for them that generate and validate blocks. This leads to fewer nodes doing calculations as several nodes can use the same representative, and this helps speed up the creation of blocks as there are fewer nodes that have to validate the new blocks. This approach also helps if something like a change to block size has to be made to the blockchain (Hongning et al., 2017).

Ripple uses an approach where collectively trusted subnetworks are created within the bigger blockchain network. Nodes here can be either servers or clients. Servers work like the main nodes of a subnet and handle the validation of new blocks, while clients only transfer funds. Servers have a unique node list that is used to determine which nodes to send a validation request to. When 80% of the nodes in the list agree that the transaction should be added, the server adds it to the chain. The blockchain is shared between all servers in the bigger network (Hongning et al., 2017).

Tendermint works in the same way as practical byzantine fault tolerance except that nodes that want to add a block to the blockchain have to lock their currency, which means that they cannot use it until they unlock it again and then they can no longer validate new blocks. It also has a three-step approach where the first step is that one of the validators wants to broadcast a new block, and two-thirds of the other validators have to confirm that the selected broadcast should happen. Then, two more votes happen where two-thirds of the nodes have to agree on each vote, and finally a block is added (Hongning et al., 2017).

5.4 PUBLIC KEY INFRASTRUCTURES

This section explains some PKI basics that might help with understanding some of the systems in the results section. To start, PKIs are cryptography solutions where devices have a public and private key pair used for encryption and decryption, respectively. The keys can be created using different cryptography algorithms depending on the security needed or the computational power available. To help devices authenticate in such a system, a certification authority (CA) has to be used that handles trusted certificates related to these keys and determines who can be trusted. However, there is an intermediary between the user devices and the CA called the registration authority (RA), where a user has to authenticate by other means than the keys before the CA will trust the device and give a certificate for the keys. The security of a system like this therefore lies with both the RA and the CA, as a failure in either could authenticate an unwanted device (Hunt, 2001).

5.5 BLOCKCHAIN AND INTERNET OF THINGS

As summarized by Conoscenti et al. (2016), many of the solutions back then used the Bitcoin blockchain, which was identified as highly unsuitable for IoT applications because of its 10-minute increments between block writes and scalability issues because of the blockchain size (this restriction is identified as "General storage and scalability issue of the blockchain" in the result section). The initial list of big problems going into this review was therefore latency, blockchain storage issues, and how to bypass PoW computational demand restrictions for devices.

The two big initial problems within IoT that the blockchain was meant to fix were scalability and security for these really heterogeneous systems (Cho et al., 2017; Deters et al., 2016). If these problems are solved, it would not only be great for the IoT itself but could potentially be used as inspiration for other heterogeneous IT systems as well.

5.6 PROBLEM DEFINITION AND MOTIVATION

As a result of the review conducted by Conoscenti et al. (2016), the Bitcoin blockchain is identified as unsuitable for IoT applications, and the need to check into other blockchains is identified. The idea arose to see what has happened since then with the attempts to match these two technologies together. With both technologies maturing and new blockchains and consensus algorithms being introduced, the possibilities should have opened up.

As mentioned in the background, the two biggest identified challenges for IoT that blockchain is meant to address are security and scalability. Seeing as this is a very broad problem, looking into this could mean that the results could be used for other technologies facing these problems as well. Decentralization while still maintaining a high level of privacy and security is, on paper, exactly what the IoT needs, especially at the rate it is expanding. But how would one go about integrating these technologies with each other? Crosby et al. (2016).

Previous research on using blockchain for IoT authentication specifically seems to be lacking. The research that can be found is on general blockchain usage together with IoT. An example of this is Deters et al. (2016) doing some testing on the general problem with a blockchain for IoT. Where to store the blockchain as edge devices have restrictions on computational power and storage availability? Some testing was done with cloud storage, but the latency of such a solution was lacklustre even with a few IoT devices, leading to the conclusion that a storage solution for the blockchain has to be found. Then, there is the research of Cho et al. (2017) that did a proof-of-concept test using smart contracts in the already established Ethereum blockchain and managed to control a light and an air conditioner from a smartphone. But it has a 12-second transaction time limit, which is not an acceptable delay for all applications. They also encountered the issue of how light should access the blockchain because it cannot store it on its own. This was the status of research 3 years ago, and the idea is that the new solutions from 2019 and beyond could have a solution to the storage problem and then find out what other problems have cropped up from these solutions being implemented.

These facts, together with the scope of this chapter, led to the research question: What are the current issues with implementing blockchain technology for decentralized authentication in IoT solutions?

The idea of this chapter is to help with future research as these technologies' working together is still, seemingly, in its infancy stage. This should be achieved by both showing proof-of-concept solutions and what needs to be solved for those systems to be put into practise or if any general problems with blockchain still have not been resolved. If a solution has been created that fixes the general problems of the blockchain technology, what should be researched next and so on?

The tie-in with the network and system administration and IT areas comes from both these technologies in the form of digital security, digital privacy, digital authentication, networking, data storage, and data integrity.

The two methods considered were doing an experiment or conducting a systematic review. During preliminary research into the topic, it became clear that several experiments had been conducted in the area within the last year, despite the fact that the searches only turned up one systematic review from 2016. With this in mind and the fact that no specific solution to a problem within this area had been identified at this time, a systematic review was deemed way more likely to give a result that is of importance to the question as stated above.

5.7 METHODOLOGY

This section describes the steps used to answer the research question at hand with included motivations for choices made. The chosen approach is to do a systematic literature review. As explained by Kitchenham and Charters (2007), the three major reasons to do a systematic literature review are:

- To summarize the gathered data for a given subject,
- Identifying gaps in research about a given subject,
- Creating a framework to guide future research about the given subject.

This chapter focuses on future research; the process of identifying what should be researched will mean that the other two reasons should be covered as well. Since the areas that need research cannot be identified without finding gaps and summarizing or at least showing what has been researched. The general guidelines for performing a systematic review have been taken from Kitchenham and Charters (2007).

Kumar, A.S. et al. (2022) explain in simpler terms that the stages of a literature review can be defined as the following steps:

1. Formulating a research question or aim,
2. Performing searches for literature,
3. Using inclusion and exclusion criteria to filter results,
4. Assessing the quality of literature,
5. Extracting data,
6. Analysing data.

5.8 SEARCH TERMS

For search terms, the keywords from the research question have been split up into "decentralized authentication," "Internet of things," and "blockchain." These are then put together in conjunction with the Boolean operators AND and OR in different variations. The terms do not allow for many variations as they are all requirements for the papers gathered and therefore the ultimate search term ended up like this: "blockchain" AND "internet of things" OR "IoT" AND "decentralized authentication".

The ultimate search terms for each database, determined from a minor prestudy, are presented in Table 5.1.

Table 5.1 Search terms per database

S. No	Database	Search term(s)
1	IEEExplore	"blockchain" AND "internet of things" OR "IoT" AND "decentralized authentication"
2	ACM Digital library	"blockchain" AND "internet of things" AND "decentralized authentication" OR "IoT" AND "blockchain" AND "decentralized authentication"
3	ScienceDirect	**Search Term 1:** "blockchain" AND "internet of things" AND "decentralized authentication" **Search Term 2:** "blockchain" AND "IoT" AND "decentralized authentication"
4	SpringerLink	**Search Term 1:** "blockchain" AND "internet of things" AND "decentralized authentication" **Search Term 2:** "blockchain" AND "IoT" AND "decentralized authentication"

Source: Author's own.

5.9 DATA EXTRACTION AND ANALYSIS

Gibbs (2007) mentions an approach to filtering out the relevant parts of the collected papers called open thematic coding. It is explained as a coding approach where the themes and categories are created as the data is collected and makes comparisons between the data.

Collected text is an example of a general idea, and the titles should be created with these general ideas in mind. This simplifies writing the data into the report as what data should be under what headline is defined during the extraction.

The comparisons also help with seeing the likeness between different extracted data from papers if they fit under an already created category.

The data being extracted should either be related to solving a problem within the adaptation of blockchain technology for IoT authentication or identify an unsolved problem within the adaptation.

5.10 SEARCHES AND PAPER ASSESSMENT

Table 5.2 presents the papers excluded after applying the inclusion and exclusion criteria.

Table 5.2 Papers left after applying the inclusion and exclusion criteria

Database	IEEExplore	ACM digital library	SpringerLink	ScienceDirect
Starting total	30	4	8	16
Title exclusion	29	3	6	10
Duplicate exclusion	29	3	3	5
Abstract exclusion	21	2	2	2
Lack of access	21	2	2	0
Final total	21	2	2	0

Source: Author's own.

5.11 CLUSTER-BASED SOLUTIONS

The most prominent theme that can be found within the collected papers is solutions based on clustering the network into smaller branches. The general idea is that the clusters themselves either have a local blockchain or no blockchain at all, and one device with the resources to handle the computational requirements for blockchains is seen as a cluster head. This cluster head is then used by all low-powered devices within the cluster it controls to perform the tasks that they cannot do. Like handling cryptography and hashing algorithms and storing data.

Aung and Tantidham (2019) have brought forward a design for use with home services or emergency services. It has a gateway device that forwards data from sensors in a house to the closest mining device. The mining devices are used to validate transactions and store the blockchain. The gateway device would send emergency calls to a miner that would validate the transaction and send it to a device owned by emergency personnel. It uses a more traditional approach of letting users login with their name and password on a website. It creates unique keys for users based on their login. It uses PKI to keep transactions encrypted during transport.

Issues: It needs testing on a larger scale. How to overcome the 12-second block write wait on the Ethereum blockchain is not yet explained. Blockchain's general storage and scalability issues apply.

From Dimopoulos et al. (2019), we show a system that is built for device-to-device communication without the need for continuous connectivity to the blockchain from either the IoT device side or the user side. MAC integrity verification is used for low-resource devices, and full asymmetric key cryptography can be used for devices that have resources for computation. The system has one server for handling client authorization requests and another server for handling IoT device authentication requests. These servers directly interact with the blockchain, while the client devices and IoT devices have to go through these servers to access the blockchain.

If one of either the user device or IoT device has constant network access, the system can be set up for the side that has continuous connectivity to handle the requests for the other, meaning a device could function without direct access to the blockchain. The connectionless side would send its authorization requests through the connected side using device-to-device communication for whatever requests are sent (Dimopoulos et al., 2019).

A second solution that works if neither the client nor the IoT device has continuous network connectivity is explained. To work, it requires that the client connects to the client authorization server at least once. It would then get the required credentials for device-to-device communication without the need for any of the devices to be connected to the blockchain during communication (Dimopoulos et al., 2019).

A solution allowing a smart contract to handle authentication is also talked about, which means real decentralization since every node containing the blockchain would handle the requests instead of just servers like in the previous two solutions (Dimopoulos et al., 2019).

Issues: There is a significant delay as a result of advanced smart contracts and the use of public blockchain. It should be tested on a private blockchain for better latency results. Blockchain's general storage and scalability issues apply.

Ali et al. (2019) describe a system where in the first phase of deployment, one device from each zone is designated as a main or master node, which can be considered as a CA. Any node can be defined as a master, but they are assigned to the node that is more resource capable and powerful, which would be recommended. All the other nodes in each zone are known as followers. Every master node creates a group ID and sends a signed ticket to each follower for identification. For the first transaction of any follower, it must require authentication. After that, an association of the follower and master is stored in the Blockchain (BC) for future needs. A local blockchain is deployed in every zone and populated with the hashes of transactions generated by IoT devices or communication between devices or nodes.

General Storage and Scalability Problems: Blockchain can help with general storage and scalability problems.

Asim et al. (2020) have created a system where each IoT system has a corresponding blockchain-enabled fog node that is closest to the system and is used for the registration and authentication of the IoT devices belonging to the same system. The devices will first register with their corresponding blockchain-enabled fog node. The identity of these devices is stored in the blockchain as transactions and blocks are created for them. These blocks are then distributed among all the other connected block-chain-enabled fog nodes. By verifying its identity credentials, blockchain-enabled fog nodes can authenticate any device belonging to a group. The blockchain validates the provided credentials, and if the credentials appear to be valid, then the devices will be successfully authenticated. Otherwise, the device will be rejected and will not be allowed to enter the network.

The proposed mechanism also provides access control for the devices in the IoT system. In the proposed system, devices can only communicate with other devices that are verified and registered with blockchain-enabled fog nodes. The device that is not registered on the blockchain cannot authenticate itself and hence cannot connect with the authenticated devices, whether they are in the same group or not. This reduces the probability of the rogue device interacting with the legal device. Moreover, the device can only register with one blockchain-enabled fog node (Asim et al., 2020).

When choosing a cryptography algorithm for the keys, the Elliptic Curve Digital Signature Algorithm (ECDSA) algorithm was chosen because it has lower power requirements while providing the same security as Rivest–Shamir–Adleman (RSA), which was the other considered algorithm. The blockchain-enabled main device with SID. Contract for initial authentication and follow-up authentication is geared towards the block created by the smart contract. The admin has, by his own intervention, put the key created by the main devices into the edge devices and put the addresses of the edge devices into the main device (so not an automatic process). Edge device authentication uses the data given to it by the admin, much like in a traditional system. Once authenticated, an auth pass is added to the edge device so it does not have to go through this process again and instead can just check itself against a created block (Asim et al., 2020).

Issues: Ethereum's 12-second block write is not addressed. It is only partially decentralized. Initial authentication is central. For functionality, it needs edge devices, which are not widely available. Blockchain's general storage and scalability issues apply. Manual addressing is not what you want for scalability.

Pajooh and Rashid (2019) developed a system that is made for a 5G connected IoT network that utilizes machine learning and intelligent clustering techniques. The system uses multiple tiers for devices, with infrastructure containing clustered IoT nodes, a core that has control devices containing local blockchains and devices that work as cluster heads, controlling their respective clusters and giving access to the local blockchain, and lastly, high-level layers that contain base stations that work as miners that the local blockchains can share and utilize. The clusters are created automatically on a geographical and available energy basis, and which device becomes the cluster head within the cluster is calculated using an algorithm for best results. Authentication within the model is split into two forms.

In a local authentication model cluster nodes work as local authentication nodes, which assigns lightweight session keys to all the devices it controls in a cluster and handles all authorization and authentication within it.

It also handles the registration of new devices into the cluster (Pajooh & Rashid, 2019).

A distributed core model that connects all the cluster heads to each other using blockchain-based communication with smart contracts distributes certificates between them for authentication. Connections for IoT devices between clusters are handled at this level, and the cluster heads set up the needed authorization and authentication for the intercluster communication to begin. With the external and internal authorization being separate, the damage from an attack on a cluster head can be limited. Since the cluster heads are selected from the IoT devices, lightweight cryptography is used in this outer network as well for compatibility (Pajooh & Rashid, 2019).

Issues: Unlike most other systems, security for a system like this is not completely solved. Is an intrusion detection system needed? Scalability issues associated with handling the huge data traffic from the cryptography exchanges during intercluster connections need to be addressed. The general storage issue of blockchain also applies.

Gountia et al. (2019b) describe a simpler system that uses gateways to handle communication between edge devices and the blockchain. Devices are identified by a public/private key pair and an Ethereum address. The admin has to couple edge devices with gateways as well as register devices with IDs manually. Edge devices can only use one gateway. A single smart contract handles all functionality like initialization of the network, registration of devices, and authentication.

Issues: Gateways are a single point of failure for clusters. Blockchain's general storage and scalability issues apply. Manual bindings are not preferred for scaling purposes.

Altun and Dalkılıç (2019) show a solution that uses the Proof of Authority consensus algorithm. They proposed a custom chain where block writing time can be set. A smart contract containing an access control list is used to authenticate nodes. A proxy node is used to relay data between edge devices and the blockchain. The edge devices have dedicated hardware to handle cryptography and hashing algorithms and power enough to create their own public and private key pairs. Each edge device is hard coded with the address of the gateway it uses and the smart contract.

Issues: Having to manually give addresses to devices is not preferred. It requires edge devices to have functionality not present in already deployed environments. Gateway is a single point of failure for the cluster it belongs to. The general storage and scalability issues of blockchain apply. Second block write time on the chain and still 13 seconds to authenticate.

Liu et al. (2019) explain a vehicular IoT solution focused on privacy. It uses roadside units, vehicles, and proxy vehicles to handle authentication. Public and private keys are used together with secret sharing to keep communication secure and authentic. Vehicles themselves authenticate together with a roadside unit to create clusters and authenticate new vehicles into the cluster.

It is not suitable because it is computationally intensive for anything that does not have edge devices with great power (basically vehicle IoT only systems). Backwards compatibility with older smart cars is not addressed. Blockchain's general storage and scalability issues apply.

Choo et al. (2019) have an IoT solution created for underwater applications. The network is separated into clusters. Each cluster is managed by a cluster head, which is assumed to have no restrictions on power or energy, unlike edge devices. Symmetric encryption is achieved with a shared key. The cluster heads handle creating and distributing the keys to devices. Devices can move between clusters. It deploys the Proof-of-Authentication consensus algorithm to cut down on the computational cost of adding blocks to the chain. Successful authentications are stored in blocks of the blockchain. This means cluster heads can check the blocks to find out if a device can be trusted without the need to go through the authentication process again, reducing overhead in the network. Edge devices in the same cluster can communicate with each other. Only cluster heads send data outside the blockchain (to the data centre).

Issues: Only simulated so far. What happens as the blockchain grows and the cluster heads run out of storage? Generally, it is just assumed that the cluster heads can do anything. Does it still work in a real scenario where cluster heads are also restricted? Initial high authentication time of 12 seconds, but with no need to reauthenticate, it is still a feasible solution for some appliances. What happens if a cluster head breaks? A single point of failure. Where to put the cluster head for other applications other than underwater?

Ahmad et al. (2020) have made what they call xDBauth. It uses global and local smart contracts. Global contracts authenticate external users, and local contracts authenticate internal users. Clusters are called "IoT domains." An IoT domain is something like a single smart home, smart hospital, etc. These domains all need a cluster head that stores a local blockchain with a smart contract to handle authentication within their domain. The cluster heads are then added to a network with a global blockchain that will be used to bind all these clusters together. Only policy hashes are stored within the blockchains, and other data has to be stored elsewhere to keep the blockchain size down. Owners of a domain will have to send an authentication request to a blockchain manager to join the global network. The edge device has to be able to handle generating its own public and private key pair and generating an ID in the form of a hash of the public key.

Issues: Assumes that edge devices are equipped with trusted platform modules (a special chip that handles encryption keys). Not true for deployed devices, and retrofit is out of the question. Blockchain storage and scalability are only partially addressed, which need a place to store all the data that is not in the blockchain, which is not explained.

Pavithran and Shaalan (2019) present a system designed to eliminate the need for a centralized CA in a PKI. It is assumed that edge devices can perform basic cryptographic functions that can store private keys. Base clusters on the location of devices. Clusters are handled by cluster heads, which are not resource-constrained. Cluster heads hold the blockchain.

Issues: Authors themselves recognize that enforcing something like this in a heterogenic environment is "a challenge". Managing keys for billions of devices is not easy. It does not handle the confidentiality of data sent between edge devices and cluster heads. Blockchain's general storage and scalability issues apply.

5.12 PUBLIC KEY INFRASTRUCTURE REWORKS

Another big theme is replacing a standard PKI with one using the blockchain as a CA instead of a centralized server. They mainly make use of smart contracts to handle the authentication and delivery of keys to each device. Some even store the keys in the smart contract itself.

Malarkodi and Satamraju (2019) came up with a solution that proposes a simple public key authentication method for IoT devices using lightweight elliptical curve cryptography. The IoT devices would hold their private keys, and public keys would be stored in the blockchain. Only devices with a valid private key can then enter the network and use the private key, which is how privacy is ensured.

Issues: How private keys are added to the IoT device is never mentioned. It mentions autonomous private key distribution can be achieved with smart contracts, but not how very generic. It assumes edge devices have storage for a private key. Blockchain's general storage and scalability issues apply.

Chamoun et al. (2019) have a system that replaces the CA server by using a smart contract to handle all things related to the public keys and authentication for the devices. It uses an Ethereum wallet for each device to store the respective private and public key pair. Devices then also get an Ethereum address.

Problems concept stage. Privacy is potentially compromised because of a traceable Ethereum address. Blockchain's general storage and scalability issues apply.

Gountia et al. (2019b) simply propose that the Ethereum blockchain can be used to create unique smart contracts for different applications and functions and have them all share the same blockchain while authenticating in different ways. It would use PKI to achieve security.

Problems: This is more of a proof of concept for blockchain and IoT than an actual solution.

Doss et al. (2019) have a vehicle-based solution that consists of three main players called Cloud, road side unit (RSU), and vehicle. The cloud contains blockchain and acts as a key distribution centre for a PKI solution. RSU acts as a middleman for communication between vehicles and the cloud. Vehicles among themselves handle the trust part of authenticating new vehicles. I need 51% of vehicles to agree to a valid join. A 51% majority is also required for a block to be written on cloud blockchain. Security-focused IT utilizes blockchain encryption for the privacy of data.

General Storage and Scalability Problems: Blockchain can help with general storage and scalability problems.

5.13 BLOCKCHAIN SERVERS

These solutions have tried to incorporate blockchain technology seemingly solely for their security benefits and uses as a single high-powered node within a network to store and handle the blockchain. Essentially making it a server just with blockchain technology for storing and handling the data.

AbuNaser and Alkhatib (2019) present a system where blocks contain two headers instead of one. The first header contains the regular hash header and a policy header. The second header contains general identification that the system needs devices with computational power and storage to accommodate the requirements. The policy header is used for settings in the blockchain, and to change settings, the policy header of the last block should be edited by a user. Edge devices authenticate at the miner and get a key and can then communicate with other authenticated devices without the need for a third party. The owner has to define how long a communication between devices can take place before reauthentication has to occur.

Issues: Not fully decentralized. Miner is a single point of failure. Uses user-defined settings to mitigate attacks. Needs users to change more things compared to a fully decentralized solution. Not a realistic scenario with things like ledger size becoming out of control having no solution and no suggestions for hardware. Assumes edge devices can hold key, not necessarily true.

Augusto et al. (2019) made a shipping and logistic-centred system. All functions and some data are stored in smart contracts on a custom-emulated Ethereum blockchain. The blockchain is permissioned which means that it's only accessible to nodes with authorization. Users get a role in this system which indicates what devices can be accessed. It uses a gateway for the edge devices that handle the mining and storing of the blockchain. Edge devices are authenticated by a key stored on them. Also, suggests keys can be saved in a wallet on the gateway.

Issues: No solution to the storage of the blockchain. The storage problem, in turn, is a problem with scalability as more data equals bigger blockchain to store. The desired approach assumes that edge devices can hold keys, not necessarily true.

Asif et al. (2020) explain an Ethereum-based Long Range (LoRaWAN) system where an edge device uses a gateway to communicate with a server that handles authentication. The gateway is only a relay in this setup. It does not handle anything else. All servers share decryption keys to keep communication encrypted no matter what server in the network an end device is connected to. PoW means that the potential delay before block is created and authentication goes through first time as well as increase computational requirements. A smart contract is used to store and receive the edge device information. The smart contract can handle the following authentications without the need for transactions being validated, meaning overhead only applies first authentication attempt.

Issues: General storage and scalability issue of blockchain applies. Uses PoW, which could consider other consensus algorithms even if the system can handle it.

Using IOTA, Wang et al. (2022) came up with a lightweight and scalable way to manage the identities of IoT devices and control access to large amounts of IoT data.

Ahmed et al. (2022) proposed the technique based on a simple and rapid consensus mechanism in which nodes are validated using their public key value and the public key value of their associated cluster head.

Issues: The framework was not flexible for IoT device migration by incorporating authentication values from the authentication table into Ethereum smart contracts.

Aghili et al. (2021) design two blockchain-enabled protocols, namely, closed-loop RFID systems (CLAB) and open-loop RFID systems (OLAB), for closed-loop and open-loop RFID-based IoT systems.

Issues: Mutual authentication protocol was missing, which can support the ownership transfer of the tags using a decentralized system

The important issue of IoT device authentication and block hash was the focus of Fotohi and Aliee (2021). The first phase of the proposed solution is to authenticate each device on the blockchain platform using an identity-based signature. Second, the device IDs were used as public keys for hashing and delivering blocks.

Using the HyperLedger fabric platform, Athavale et al. (2020) conclude that IoT data may be securely stored, validated, and authenticated using built-in encryption and signature mechanisms. Transport Layer Security (TLS) encryption ensures that data cannot be accessed during internal network communications in HyperLedger.

5.14 ADAPTATIONS OF THE BLOCKCHAIN TECHNOLOGY

These solutions have tried to change something about the blockchain technology itself, like how the blockchain works or how a consensus algorithm works. One of them is even included in other solutions presented in this paper, so this seems to be a great area of interest to actually combat the problems of blockchain technology.

Du et al. (2019) describe a new type of blockchain called "redactable consortium blockchain." It allows for changes in the blockchain, which is useful if errors are entered into the chain or in general, if data has to be changed for whatever reason. They proposed a new hash and signature as building blocks for this new type of blockchain designed specifically for usage with power-restricted IoT devices. In this approach, a chain network is created by a chain manager. It arranges memberships for all nodes and runs smart contracts to create a network. Once a block is added to the blockchain, the chain manager is no longer needed and gives ownership of the blockchain to the authorized nodes and goes offline. A user sensor is a device that uses a newly developed algorithm to authenticate transactions and give authorized nodes the go ahead to update signatures. Authorized nodes are allowed to write and redact on the blockchain by all working together. Redaction can be done without creating forks of the chain, and signatures can be updated during the redaction without the need for the user sensor. Judge sensors are there in case disputes happen during transactions and can differentiate between signatures from user sensors and authorized nodes. Nodes share a key used for authentication to avoid the need for calculations to authenticate. New hashing algorithms for blocks that are better optimized for use with low-power edge devices. It has functionality that allows for the redaction of blocks based on all participating nodes' trustworthiness thanks to blockchain security. It uses smart contracts to handle authentication requests.

General Storage and Scalability Problems: Blockchain can help with general storage and scalability problems.

Das et al. (2019) have created a new lightweight consensus algorithm called Proof-of-Authentication designed for resource-constrained devices. It still follows the traditional concept of only updating when a new block is created. All nodes can create transactions and blocks, but the validation of new blocks has to be validated by trusted nodes. Each of the nodes has a trust value in this system. An untrusted node has a value of 0, and a trusted node has a value of 1. Every validation a trusted node does increases the node's trust value, much like how nodes get currency in PoW cryptocurrency blockchains. The higher the trust value, the more likely nodes are to be chosen to validate. It uses public key cryptography for authentication and block validation.

Issues: We just need more testing and potential security considerations with implementation.

Kim (2020) proposed solving the PoW problem by using a reverse sequence hashing algorithm. A physical system consists of a creator node that creates and holds a private blockchain and also mines the new blocks. It has a manager node that handles authentication and authorization of edge devices. Edge devices are called member devices and have to be invited to the network by a manager node. Only when a block is written that shows that a member has been authenticated, can that member start accessing resources on the network.

Issues: only the concept. It needs to be further studied for practical application.

Brandão (2020) proposes a system for industrial control systems (ICS). It has the unique solution of having a maximum set blockchain size and what would be considered a rolling blockchain where the first blocks are removed while new ones are appended. The only nonedge device in the system is called a "pub" and is only used to initially authenticate an edge device in the system. The pub has a predefined list of codes that the admin of the system has created. To authenticate an edge device, they would have to send one of these codes from the list to the pub for authentication. Once authenticated, the edge device is added to a list the pub keeps of the authenticated devices. The blockchain is then synced from all pubs the edge device is associated with. Each node needs to be able to hold a pair of private and public keys used for identification, authentication, and encrypting of messages. They also need to be able to hold the blockchain as most of the communication is device-to-device. At least one message needs to be kept in the blockchain at all times, as the standard new block points to the header of the old block system still applies.

It needs specialized edge devices with enough storage to fit the max size set for the blockchain not fit for already established systems. The author identifies the private key as a potential security risk.

5.15 DEVICE-TO-DEVICE

These systems are mainly based on edge devices. They themselves handle all the business without the need for any centralized servers or nodes.

Anand et al. (2019) describe a system where a secret key in the shape of a nonstatic integer value only known by the two devices communicating is used. The integer value changes between every message. A hash of the message is sent from the start node to an intermediary node that then forwards the hash to the destination. This hash is then used by the end node to check the message received from the start node to help data integrity and prove that the secret key is the right one. It uses its own currency, called "mirage tokens." Devices have to spend one token to send a message to another device. Tokens are earned by acting as an intermediary node (2 per time).

Issues: Recognizes the shortcomings of the initial network, which had a lot of overhead as devices tried to earn tokens. When established, this overhead is removed. It does not mention what hardware is expected of nodes (it seems like they need more power than most would realistically have). Only simulated. It only mentions how the messages and the secret key could be encrypted. It is not a usable solution in the present scenario.

5.16 MANAGEMENT

Gountia et al. (2019a) present a solution called DecAuth that would work like a "login with third party account" button on sites. User authentication is tied to a wallet, much like the already existing cryptocurrency solutions. A web page is used to communicate with a smart contract within the blockchain that allows the handling of the wallets like creation, changing keys, or deletion. This allows for decentralized credentials with high availability.

Users will have to handle keys and login with them. While the computer side has better security, the same security issues as any other login on the human error side exist. Only addresses authentication for managing a system, not the system itself.

5.17 SUMMARY OF THE IDENTIFIED PROBLEMS

As seen in Table 5.3, the most common problem by a large margin is that the creators of the systems have ignored the problem with a growing blockchain and how that should be addressed for using the system over a longer time. The second biggest problem is that the systems would need specialized devices. This does not have to be a problem for completely new systems, but for integration with anything that is already deployed, this is a deal breaker, and it is therefore a big deal. Some of the systems were not tested enough or tested at all in practice. The scalability issues mentioned do not have anything to do with the storage issue of the blockchain; instead, they are systems that have further scalability issues by needing user input or preset device lists and things like that. Latency issues are still around for some systems; some systems have single points of failure and other nondecentralized features and some have security problems.

Table 5.3 Summary of problems identified from the results

S. No	Problems identified	Proposed solutions
1	General storage and scalability issue of blockchain	17
2	Specialized edge devices	9
3	Needs more testing	6
4	Scalability problems	5
5	Latency	4
6	Not fully decentralized	4
7	Security problems	4

Source: Author's own.

5.18 DISCUSSION

The general result shows that some of the limitations of the blockchains are still problems when implementing these technologies together. However, from a purely authentication standpoint, a system that uses smart contracts in the Ethereum blockchain could be used to authenticate without latency as the smart contracts can be called at any time. The main problem is that if any authentication data is also supposed to be saved to the blockchain, the 12-second interval between blocks being written comes into play. It is still hard to see how a profit-seeking entity would ever adopt a system where each authentication attempt would have to wait 12 seconds between tries or confirmation. So, while smart contracts are a great asset, the storage on the Ethereum blockchain would have to be addressed. To do this, a new blockchain would have to be created. Smart contracts are an example of what the blockchain needs more of to become a widely adoptable technology.

Most systems use some sort of gateway between the blockchain and the edge devices, which acts as a single point of failure, which defeats the whole decentralization aspect of implementing the blockchain technology in the first place. Some of them have solved it by storing the blockchain on these gateways and letting them communicate with each other. But they all have some other problem, like edge devices needing cryptography capabilities, which is untrue for systems already put up. In general, there is no single solution that pops out that seems to have solved the shortcomings of these technologies. All of them have a hurdle to get over, and it is surprisingly rarely the same one. There is not a single system presented that has been tested on a long-term basis and does not have any drawbacks or problems.

The majority of the solutions have focused on the security benefits of the systems. The problems of integrating blockchain technology are skimmed over.

Several systems protect against a similar list of attacks:

Replay Attacks: when a rogue device tries to resend a valid data transmission to appear to be a valid device in the system.

Man-in-the-Middle Attacks: they occur when a rogue device intercepts a communication between two points in the system in order to listen to the data sent.

Denial of Service Attacks: in which the system is flooded with useless data in order to prevent useful data from entering.

Sybil Attacks: they are attempts by an attacker to take over the system by creating a large number of identities and becoming the majority.

They use two-way authentications. So, while all problems with merging these two technologies still are not solved, some real security benefits have been presented in these partial blockchain solutions. This could mean that a third technology could need to be introduced to this mix to achieve the full potential of the proposed solutions.

From a purely authentication-based stance, almost all of these systems do a really good job of creating a secure way of authenticating all the devices. But general assumptions made from a big chunk of them with things like edge devices having more power than most actually have or having cryptography specific chips embedded are just not a feasible solution except for newly created systems. With the number of devices already in circulation, spending time researching such a system is essentially wasted as retrofitting all devices is not an actual solution and probably impossible for some devices.

The main area where these solutions could have a chance to be implemented would be vehicular IoT, as vehicles have more room and power enough to handle a computer strong enough to keep up with the demands of blockchain technology. However, this would mean that every car manufacturer from now on would have to agree on a standardized solution to ensure that the computers have the right functionality. Retrofitting something like this to old vehicles is not feasible. The real conclusion to draw from all this discussion is that unless something big happens within the research of these two technologies, no realistic scenario exists where they become widely adopted.

5.19 CONCLUSION

The systematic literature review resulted in twenty-five articles. They were then analysed and summarized. The result, in general, shows that most of the proposed systems need some sort of specialization of the hardware in the IoT system to fully make use of their functionality and that most systems do not take the shortcomings of the blockchain technology regarding scalability and storage into consideration when presenting their solutions. It is clear that these technologies working together are still in a very early stage.

Some benefits on the security front are identified, and for new systems that can use hardware that is specialized to handle the extra overhead, the blockchain technology currently adds one of these solutions can be considered. But for the general scope of IoT, the presented solutions do not handle backwards compatibility good enough without creating new hard to solve problems that cannot be seen as the next big step for IoT. None of the gathered solutions present a problem-free one-size-fits-all type system that could be easily migrated to which would be necessary for wide-scale adaptation from profit-seeking entities.

The blockchain technology does not seem mature or developed enough for this application at the time of writing. At this time, to become some sort of standard that would benefit a decentralized solution like blockchain, the solutions would need to be clear upgrades without drawbacks over other possible solutions within the IoT area, and at this time, that is not achieved. The main problems that still have to be addressed are scalability and storage problems related to blockchain technology itself. The other issue that could potentially be looked into is cryptography, where the need for private key storage on edge devices is a problem holding back some of these systems from being implemented into already existing systems. Perhaps even replacing PKIs with something else entirely within those systems would allow current edge devices to exist in a blockchain-based system.

5.20 FUTURE WORK

The general feel of almost all these solutions is that they are trying to solve a problem as best they can with the tools that are available. The point they are missing is that they do not have the right tools for the job. If these technologies working together are to become a standard, more work equivalent to that presented in section 5.5 has to be done. Were the basics and functionality of the blockchain and consensus algorithms changed and adapted at a core level? Otherwise, none of these systems will reach their full secure and decentralized potential.

REFERENCES

AbuNaser, M., & Alkhatib, A.A.A. (2019). Advanced survey of blockchain for the internet of things smart home. *2019 IEEE Jordan International Joint Conference on Electrical Engineering and Information Technology (JEEIT)*, 58–62. DOI: 10.1109/JEEIT.2019.8717441.

Aghili, S.F., Mala, H., Schindelhauer, C., Shojafar, M., & Tafazolli, R. (2021). Closed-loop and open-loop authentication protocols for blockchain-based IoT systems. *Information Processing & Management*, 58(4). DOI: 10.1016/j.ipm.2021.102568.

Ahmad, N., Ali, G., Azaz, A., Cao, Y., Cruickshank, H., Khan, S., & Qazi, E.A. (2020). xDBAuth: Blockchain based cross domain authentication and authorization framework for internet of things. *IEEE Access*, 8(2020), 58800–58816. DOI: 10.1109/ACCESS.2020.2982542.

Ahmed, M.T.A., Hashim, F., Hashim, S.J., & Abdullah, A. (2022). Hierarchical block-chain structure for node authentication in IoT networks. *Egyptian Informatics Journal*. DOI: 10.1016/j.eij.2022.02.005.

Ali, J., Ali, T., Alsaawy, Y., Khalid, A., & Musa, S. (2019). Blockchain-based smart-IoT trust zone measurement architecture. *COINS'19: Proceedings of the International Conference on Omni-Layer Intelligent Systems*, 152–157. DOI: 10.1145/3312614.3312646.

Altun, A., & Dalkılıç, G. (2019). Blockchain based confidential communication and authorization model for IoT devices. *2019 Innovations in Intelligent Systems and Applications Conference (ASYU)*. DOI: 10.1109/ASYU48272.2019.8946360.

Anand, G., Anand, S., Bashir, A.K., Chauhdary, S.H., & Raja, G. (2019). Mirage: A protocol for decentralized and secured communication of IoT devices. *2019 IEEE 10th Annual Ubiquitous Computing, Electronics & Mobile Communication Conference (UEMCON)*, 1074–1081. DOI: 10.1109/UEMCON47517.2019.8993110.

Asif, W., Danish, S.M, Lestas, M., Rajarajan, M., Qureshi, H.K., & Zhang, K. (2020). Securing the LoRaWAN join procedure using blockchains. *Cluster Computing*, 2020. DOI: 10.1007/s10586-020-03064-8.

Asim, M., Baker, T., Hung, P.C.K., Khalid, U., Rafferty, L., & Tariq, M.A. (2020). A decentralized lightweight blockchain-based authentication mechanism for IoT systems. *Cluster Computing*, 2020. DOI: 10.1007/s10586-020-03058-6.

Athavale, V.A., Bansal, A., Nalajala, S., Aurelia, S. (2020). Integration of blockchain and IoT for data storage and management. *Materials Today: Proceedings*. DOI: 10.1016/j.matpr.2020.09.643.

Augusto, L., Costa, R., Ferreira, J., & Jardim-Gonçalves, R. (2019). An application of ethereum smart contracts and IoT to logistics. *2019 International Young Engineers Forum (YEF-ECE)*, 1–7. DOI: 10.1109/YEF-ECE.2019.8740823.

Aung, Y.N., & Tantidham, T. (2019). Emergency service for smart home system using ethereum blockchain: System and architecture. *2019 IEEE International Conference on Pervasive Computing and Communications Workshops (PerCom Workshops)*, 888–893. DOI: 10.1109/PERCOMW.2019.8730816.

Brandão, R. (2020). A blockchain-based protocol for message exchange in a ICS network. *SAC '20: Proceedings of the 35th Annual ACM Symposium on Applied Computing*, 357–360. DOI: 10.1145/3341105.3374231.

Chamoun, M., El-Hajj, M., Fadlallah, A., & Serrhrouchni, A. (2019). Ethereum for secure authentication of IoT using pre-shared keys (PSKs). *2019 International Conference on Wireless Networks and Mobile Communications (WINCOM)*, 1–7. DOI: 10.1109/WINCOM47513.2019.8942487.

Cho, S., Huh, S., & Kim, S. (2017). Managing IoT devices using blockchain platform. *2017 19th International Conference on Advanced Communication Technology (ICACT)*, 464–467. DOI: 10.23919/ICACT.2017.7890132.

Choo, K.K.R., Dehghantanha, A., Parizi, R.M., Srivastava, G., & Yazdinejad, A. (2019). Energy efficient decentralized authentication in internet of underwater things using blockchain. *2019 IEEE Globecom Workshops (GC Wkshps)*. DOI: 10.1109/GCWkshps45667.2019.9024475.

Conoscenti, M., De Martin, J. C., & Vetrò, A. (2016). Blockchain for the Internet of Things: A systematic literature review. *2016 IEEE/ACS 13th International Conference of Computer Systems and Applications (AICCSA)*, 1–6. DOI: 10.1109/AICCSA.2016.7945805.

Crosby, M., Pattanayak, P., & Verma, S. (2016). Blockchain technology: Beyond bitcoin. *Applied Innovation Review*, 2, 6–19. Retrieved from https://j2-capital.com/wp-content/uploads/2017/11/AIR-2016-Blockchain.pdf.

Das, G., Kougianos, E., Mohanty, S.P., Nanda, P., & Puthal, D. (2019). Proof-of-authentication for scalable blockchain in resource-constrained distributed systems. *2019 IEEE International Conference on Consumer Electronics (ICCE)*. DOI: 10.1109/ICCE.2019.8662009.

Deters, R., Jamsrandorj, U., Samaniego, M. (2016). Blockchain as a service for IoT. *2016 IEEE International Conference on Internet of Things*, 433–436. DOI: 10.1109/iThings-GreenCom-CPSCom-SmartData.2016.102.

Dimopoulos, D., Fotiou, N., Polyzos, G.C., Siris, V.A., & Voulgaris, S. (2019). Trusted D2D-based IoT resource access using smart contracts. *2019 IEEE 20th International Symposium on "A World of Wireless, Mobile and Multimedia Networks" (WoWMoM)*. DOI: 10.1109/WoWMoM.2019.8793041.

Doss, R., Jiang, F., Patterson, N., Wang, X., & Zeng, P. (2019). An improved authentication scheme for internet of vehicles based on blockchain technology. *IEEE Access*, 7, 45061–45072. DOI: 10.1109/ACCESS.2019.2909004.

Du, X., Guizani, M., Huang, K., Mu, Y., Rezaeibagha, F., Wang, X., Xia, Q., Yang, G., & Zhang, X. (2019). Building redactable consortium blockchain for industrial internet-of-things. *IEEE Transactions on Industrial Informatics*, 15(6), 3670–3679. DOI: 10.1109/TII.2019.2901011.

Fotohi, R., & Aliee, F.S. (2021). Securing communication between things using blockchain technology based on authentication and SHA-256 to improving scalability in large-scale IoT. *Computer Networks*, 197. DOI: 10.1016/j.comnet.2021.108331.

Gibbs, G.R. (2007). Thematic coding and categorizing. *Analyzing Qualitative Data*, 38–55. DOI: 10.4135/9781849208574.n4.

Gountia, D., Jena, D., Mohanta, B.K., Panda, S.S, Patel, S., & Sahoo, A. (2019a). *DecAuth: Decentralized Authentication Scheme for IoT Device Using Ethereum Blockchain. TENCON 2019-2019 IEEE Region 10 Conference (TENCON)*, 558–563. DOI: 10.1109/TENCON.2019.8929720.

Gountia, D., Jena, D., Mohanta, B.K., Panda, S.S., & Satapathy, U. (2019b). A blockchain based decentralized authentication framework for resource constrained IoT devices. *2019 10th International Conference on Computing, Communication and Networking Technologies (ICCCNT)*. DOI: 10.1109/ICCCNT45670.2019.8944637.

Hongning, D., Huaimin, W., Shaoan, X., Xiangping, C., & Zibin, Z. (2017). An overview of blockchain technology: Architecture, consensus, and future trends. *2017 IEEE 6th International Congress on Big Data*, 557–564. DOI: 10.1109/BigDataCongress.2017.85.

Hunt, R. (2001). PKI and digital certification infrastructure. *Ninth IEEE International Conference on Networks*, ICON 2001, Bangkok, pp. 234–239, DOI: 10.1109/ICON.2001.962346.

Jayavadhana, G., Marimuthu, P., Rajkumar, B., & Slaven, M. (2013). Internet of things (IoT): A vision, architectural elements, and future directions. *Future generation Computer Systems*, 29(7), 1645–1660. DOI: 10.1016/j.future.2013.01.010.

Kim, D. (2020). A reverse sequence hash chain-based access control for a smart home system. *2020 IEEE International Conference on Consumer Electronics (ICCE)*. DOI: 10.1109/ICCE46568.2020.9043090.

Kitchenham, B., & Charters, S. (2007). *Guidelines for Performing Systematic Literature Reviews in Software Engineering*. Retrieved from https://www.elsevier.com/__data/promis_misc/525444systematicreviewsguide.pdf.

Kumar, K. S., Nassa, V.K., Uike, D., Kalra, A., Sahu, A.K., Athavale, V.A., Saravanan, V. (2022). A Comparative Analysis of Blockchain in Enhancing the Drug Traceability in Edible Foods Using Multiple Regression Analysis. Journal of Food Quality. Article ID 1689913, 6 pages. https://doi.org/10.1155/2022/1689913.

Liu, H., Zhang, P., & Zhang, Y. (2019). Blockchain enabled cooperative authentication with data traceability in vehicular edge computing. *2019 Computing, Communications and IoT Applications (ComComAp)*, 299–304. DOI: 10.1109/ComComAp46287.2019.9018754.

Malarkodi, B., & Satamraju, K.P. (2019). A secured and authenticated internet of things model using blockchain architecture. *2019 TEQIP III Sponsored International Conference on Microwave Integrated Circuits, Photonics and Wireless Networks (IMICPW)*, 19–23. DOI: 10.1109/IMICPW.2019.8933275.

Pajooh, H.H., & Rashid, M.A. (2019). A security framework for IoT authentication and authorization based on blockchain technology. *2019 18th IEEE International Conference on Trust, Security and Privacy in Computing and Communications/13th IEEE International Conference on Big Data Science and Engineering (TrustCom/BigDataSE)*, 264–271. DOI: 10.1109/TrustCom/BigDataSE.2019.00043.

Pavithran, D., & Shaalan, K. (2019). Towards creating public key authentication for IoT blockchain. *2019 Sixth HCT Information Technology Trends (ITT)*, 110–114. DOI: 10.1109/ITT48889.2019.9075105.

Pilkington, M. (2016). *Blockchain Technology: Principles and Applications*. Retrieved from https://pdfs.semanticscholar.org/e31c/a71621e1402a46ac-2c1afb2eba9a7061d139.pdf.

Wang, S., Li, H., Chen, J., Wang, J., & Deng, Y. (2022). DAG blockchain-based lightweight authentication and authorization scheme for IoT devices. *Journal of Information Security and Applications*, 66. DOI: 10.1016/j.jisa.2022.103134.

Chapter 6

Blockchain – Foundational infrastructure for Web 3.0 and cryptoassets

Ashutosh Dubey and Deepnarayan Tiwari
National Payment Corporation of India

Anjali Tiwari
Chitkara University

CONTENTS

6.1 INTRODUCTION

Blockchain and distributed ledger technologies (DLTs) are the preferred technology used to deliver cryptoassets of multiple forms and functions for various applications. DLT is based on a decentralized database managed by various servers across participants and computing nodes. Transactions are recorded with an immutable hash generated from a cryptographic signature, then grouped in blocks with append-only mode and formed a secure chain of blocks by including the previous block's hash. This DLT technology eliminates costly reconciliation settlements and automates business processes, lowering business costs and opening new business opportunities.

Essentially, blockchain technology is used to support Web 3.0 principles as a secure, open, protocol-driven, token-based platform. It is imperative to understand cryptoassets' types and features to support specific applications, since they are key to driving the Web 3.0 economy. Blockchain/ DLT has been hailed as the genesis use case for cryptoassets since they facilitate multiparty movement of value with on-demand ledger records, which facilitate efficiency and cost savings by removing friction and intermediaries. The Web 3.0 token enables decentralized Internet protocols for multiparty communication, ensuring value for computing, bandwidth, storage, identification, and hosting companies. Various innovations have also been introduced in recent years through smart contracts and consensus mechanisms, including programmability of workflows, governance validation, and access. Later, we discussed the role of blockchain in financial applications and how decentralized finance "DeFi" or the future of finance will be driven by it. Each technology has its own risks. Blockchain also faces challenges like scaling, security, and sanity in terms of decentralization. Multiple studies indicate that current DeFi applications run with some decentralization but not fully.

Our book chapter offers an overview of blockchain technology, its role in current financial landscapes, the emergence of DeFi platform, and a packaged offering for Web 3.0 applications. Blockchain emerged as one of the best applications for DLT. It is essential to understand the evolution, principles, and concepts of Web 3.0 before understanding the blockchain application. In Web 3.0, there is no centralized authority to control the network, but rather a distributed architecture and open access flavors.

In this chapter, we present a background of evaluation of blockchain technology that makes the grounds for the next frontier of Internet–Web 3.0. We also detailed the blockchain as foundation infrastructure and discussed the prospects for the blockchain in terms of what businesses and individuals can do to embrace blockchain technology successfully. Next, we discuss the importance of blockchain-driven assets with their classification. Finally, we discuss the prospects of blockchain as the future of finance and its associated risk.

6.2 BACKGROUND

We live in a world that is experiencing a digital transformation (Iivari et al., 2020). Digitalization of the services sector has been greatly facilitated by the advent of DLTs such as blockchains. Artificial intelligence (AI) and machine learning (ML) are also assisting humans in improving their efficiency at work and quality of life due to the availability of large volumes of big data. United Nations has set 17 Sustainable Development Goals (SDGs) with considerable resources devoted to achieving them. Various parties and groups worldwide are involved in SDG projects, including governments, aid agencies, local people, and NGOs. The Decentralized Consortium Funding (Aysan et al., 2021) model was recently developed by Boston Consulting Group, Input Output, Blockchain Research Institute, University of Wyoming to address the need for effective implementation of SDGs globally. This model provides a new framework for engagement, execution, and governance using blockchain as a core technology. Blockchain is built upon DLT, allowing multiple copies of the same data to be distributed across several nodes concurrently through a peer-to-peer network. Blockchain eliminates the need for centralized data storage or clearinghouses with enhanced data security.

The COVID-19 pandemic has accelerated this digitalization trend of financial markets and economies. The pandemic leads to many experimentations – one such experimentation is the use of blockchain technology to support financial assets as representations in a digital format and stored on distributed ledgers (Adrian & Mancini-Griffoli, 2021). Tokenization is a critical component of the upcoming blockchain technology's capabilities. Tokens based on DLT have been used in financial markets by means of initial coin offerings (ICOs) started from 2017 to 18 (Momtaz, 2020), although innovative technologies in finance pose systemic risks as well as technological, financial, regulatory and ecological challenges (Reiff, 2021). Bitcoin and Ethereum are two major cryptocurrencies and blockchain platforms covering 80%–85% of blockchain applications. They require large amounts of electricity and more than the total consumption by an entire country—to perform the computations and mining required for transaction processing and consensus establishment.

It is important to note that in order to achieve consensus, miners can participate in two types of mining processes – public (openly accessible by anyone), private (open for limited use case) or permissioned (controlled by a limited set of entities) DLT platforms. Developers build applications on platforms to solve business problems. The major platforms are Ethereum (permission-less, purely public), Hyperledger Fabric (permissioned, private), and R3 Corda (permissioned, private). Investors may be most interested in the decentralized applications ("Dapps") running on blockchain platforms, either as new business models disrupting existing processes or future investment opportunities.

The Bitcoin Electricity Consumption Index (CBECI) of the University of Cambridge (Rauchs et al., 2020) estimates that the Bitcoin network consumes 129 terawatt-hours of electricity per year. For cryptoassetization, new methods to reduce energy requirements are being developed, like delegated proof of stake, practical byzantine fault tolerance, proof of authority, etc., for ensuring the legitimacy of cryptocurrency transactions.

Other challenges such as data privacy and governance pose threats in the financial ecosystem around blockchain-based assets. Also, investor protection, cyber security attacks and fraud, and vulnerabilities that affect global financial stability, as well as regulatory oversight issues more akin to AML and terror funding, are a few of the features of these threats. Governments, central banks, financial regulators, and policymakers are trying to strike a balance between threats and innovations associated with the use of DLTs to provide financial services.

Decentralized finance (DeFi), an application of Web 3.0, is leveraging blockchain as the core technology to enable efficient, inclusive, and competitive financial markets. Fortune business insights project that the global blockchain market will grow from $4.68 billion in 2021 to $104.19 billion in 2028 at a compound annual growth rate of 55.8% (Blockchain, 2022). In light of stablecoins and blockchain technology's nascent nature, many jurisdictions have established regulatory "sandboxes" so that companies can test their offerings and innovate in a controlled environment with minimal regulatory requirements. The benefits of sandboxes are twofold. First, the companies can use them to understand better how their services will work. Second, in contrast, the regulators can use them to identify gaps and problems in current regulations and any new regulatory concerns that may arise. This will lead to further adoption of blockchain, and it will move from being a technology to the default choice of platform and future of business.

6.3 THE NEXT FRONTIER OF THE INTERNET – WEB 3.0

The Static Web, also called Web 1.0, was the first and most reliable Internet of the 1990s, despite offering little information and no user interaction (Rudman & Bruwer, 2016). User pages and even comments on articles were not available in Web 1.0. There were no algorithms to sift through web pages in Web 1.0, so users had difficulty finding relevant information. A one-way highway with a narrow path where the content was created mainly by a select few and information was primarily gathered from directories. Web 2.0 gave us a world of interactive experiences and user-generated content that made it possible for companies like Ola, Flipkart, OYO, Facebook, and Instagram to grow and thrive. Three core layers of innovation drove the rise of Web 2.0: mobile, social, and cloud. Even as the Web 2.0 (Rudman & Bruwer, 2016) wave bears fruit, the next significant paradigm shift in Internet applications shows signs of growth, which can be termed Web 3.0. Web 3.0 networks claim to enable distributed users and machines to interact with data, value, and other counterparties as if there were no third parties in the connection. This is a computing fabric that keeps privacy in mind while being human-centric. The Web 3.0 (Zarrin et al., 2019) platform

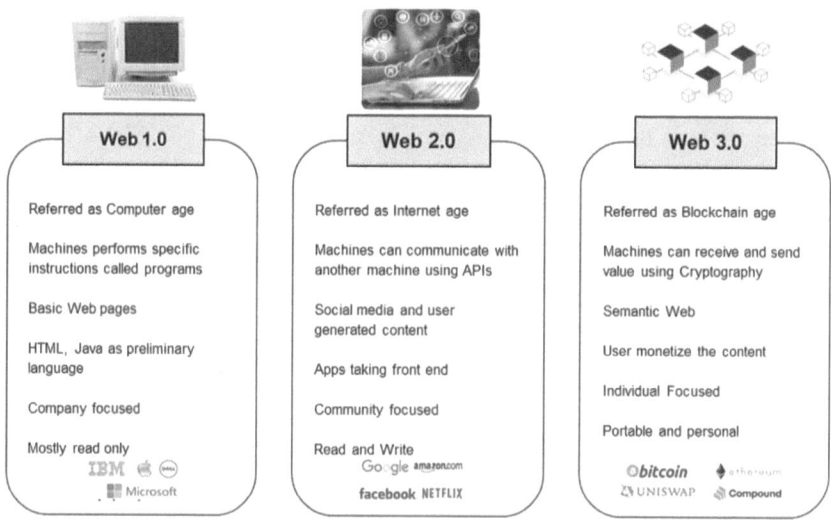

Figure 6.1 Emergence of Web 3.0.

marks the beginning of an age of open, trustless, and permission-less networks. Open-source software is developed by a community of open-source developers and implemented publicly, along with being available for open review by anyone. It is a "trustless" network in the sense that participants are able to interact publicly or privately without knowing about the involvement of third parties. It is also "permission-less," as any user or supplier can participate without authorization. A decentralized Internet would run in the third generation. Smart contracts, seamless integration, and censorship-resistant storage of user data files characterize the third generation of the web. As a result, blockchain will serve as a significant driving force for the next generation of the Internet, which drives the interaction of states between interconnected peers of the network (Figure 6.1).

6.4 BLOCKCHAIN AS FOUNDATIONAL INFRASTRUCTURE AND THEIR USECASES

The technology of blockchain (Nakamoto, 2008) is a method for recording information in a way that prevents others from changing, hacking, or cheating the system. Blockchain records transactions based on an immutable cryptographic signature known as a hash. The blockchain consists of blocks, and every time the blockchain completes a new transaction, every member's ledger is updated. One of the most popular examples of using blockchain technology is the cryptocurrency Bitcoin, which is intrinsically tied to it. Also, it is the most controversial option since it enables the creation of a multibillion-dollar market of anonymous transactions without government control (Banerjee et al., 2018). Therefore, blockchain-based firms will have to deal with several regulatory issues involving national governments and financial institutions.

Even though Bitcoin was designed as an alias for public blockchain to operate democratically and free from interference by banks or regulators, blockchain technology can still operate within closed parameters.

By implementing their blockchain, a company or bank could control which transactions appear on the chain (Wild et al., 2018). As with traditional banking, it will still be a secure system but based mostly on the trust of the decision-maker.

In addition to Ripple and R3's Corda, several other companies offer their versions of blockchain technology, including Hyperledger Fabric, Corda, and Hyperledger, which are permissioned systems (Chowdhury et al., 2018), restrict access to transaction data to the parties involved instead of making that data public as Bitcoin does (Figure 6.2). Here is the evolution of Blockchain.

In addition to the high energy requirements to maintain blockchain security, two potential solutions exist to make it a green infrastructure. One is carbon offsetting (Alvarez & Argente, 2020). For example, organizations that use crypto tokens to acquire clean energy may be able to demonstrate that their business practices are climate neutral. Using less energy-intensive blockchain technology such as proof-of-stake rather than proof-of-work is also an option that can reduce the amount of energy required by the system. It may become possible to save the environment through blockchain technology when cryptocurrencies become green. Blockchain platforms are

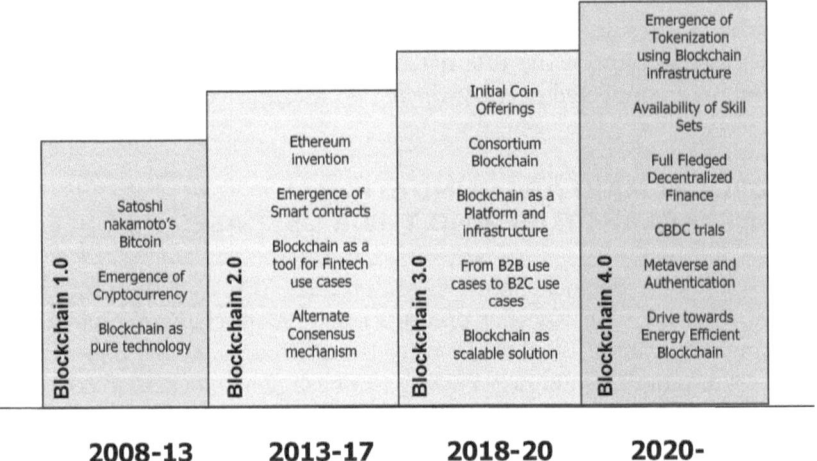

Figure 6.2 Evolution of blockchain over time from 2008 onwards.

another emerging trend for issuing currency. Bitcoin was adopted as a legal tender for the first time by El Salvador in 2021; several other distressed nations are expected to follow. Another asset class called "Cryptoassets" is emerging using blockchain as infrastructure for creation, transfer, reconciliation, and record keeping securely in a nonrepudiated manner.

The Bank for International Settlements (Chen & Bellavitis, 2020) conducted a study which indicated that more than 80% of institutions have actively developed proofs of concept using blockchain technology in order to establish Central Bank Digital Currencies (CBDCs). CBDC is a claim against the central bank as legal tender and not backed by a commercial bank or third-party payment processor. A digital ledger (which may be a blockchain) is used by the CBDC in order to secure payments between banks, institutions, and individuals.

Blockchain technology can revolutionize finance by making the market permission-less and accessible to everyone. Moreover, finance using blockchain as technology (decentralized finance or DeFi) offers the concept of composability, which means that any existing offering can be mixed and matched to create a custom solution. Composable components make such a network possible to accommodate innovations and needs in the finance sector, and smart contracts can control everything. When a specific event occurs, smart contracts automatically execute an action. In this way, rules governed by technology can be defined. A smart contract can define conditions that, if met, trigger other actions such as sending or receiving funds or even executing other smart contracts. Automated services enable existing financial services to be delivered over blockchain networks and new services to be created with rules and conditions of execution guaranteed by the network itself (Table 6.1).

Table 6.1 Application of blockchain

#	Usecase	Description
1	Instant settlement for domestic payments	Even though domestic payments are fast for end users, with credits of funds being transferred almost instantly, they remain hampered with numerous challenges. Some of these challenges include (i) inefficiencies on the back end; (ii) costs; and (iii) scalability. Most reconciliations are done between and within platforms since messaging does not facilitate the movement of values or instant settlement. Due to this, there are significant delays between the receipt of fund instructions and interbank settlement. Using blockchain and DLT as potential reference architectures and infrastructure for domestic payments (Allende et al., n.d.) has been a critical use case explored. Use cases are classified into two categories: Orchestration and Clearing In a blockchain-based payment network, the transaction can be executed and settled almost instantly, reducing the risk of delays related to payment settlement

(Continued)

Table 6.1 (Continued) Application of blockchain

#	Usecase	Description
		Instant settlement using Cryptoassets Central banks are exploring blockchain/DLT as the underlying infrastructure and technology for instant settlement and regulated open finance as part of the retail ecosystem. As a result, government-backed tokens have been expected to eliminate clearing, encourage instant settlement similar to the use of currency, and encourage more entrants to the payment ecosystem while operating within the regulatory framework and using an instrument that remains risk-free.
2	Cross-border payments	Blockchain and distributed ledger technology allow individuals and institutions to send and receive payments with minimal involvement from middlemen (Kulk & Plompen, 2021). The use of robust payment systems and/or regulated tokens supported by blockchain networks can enable the settlement of cross-border payments quickly and cost effectively and can provide the foundation for a regulated open finance ecosystem.
3	Programmable subsidy distribution	By leveraging the smart contract features of blockchain and DLTs, central banks and governments are exploring programmable tokens (Gonzalez, 2019). Tokens can be programmed to behave in specific ways. Furthermore, as a payment instrument, Tokens may have embedded executable code that can execute business rules upon meeting defined conditions set by the issuer (government) or the distributors (financial institutions) meeting specific purposes.
4	Lending and borrowing	An application can be developed that makes it possible for borrowers and lenders to directly negotiate on interest rates, payments, and credit duration through blockchain and DLT infrastructure (Chang et al., 2020). Borrowers and lenders can negotiate terms using smart contracts. Smart contracts can be linked to legal text including how loans can be used, repaid, and how disputes can be resolved.
5	Trade finance	There are currently multiple participants along the lifecycle of a trade finance transaction, making it an opaque and fragmented industry. Various companies – banks, exporters, importers, insurers, and export credit agencies – are burdened by antiquated processes. As costs and operational risks rise, the industry has utilized blockchain and DLT to simplify, better manage, and digitize trade (Houben & Snyers, 2020). Since 2017, blockchain -based trading and supply chain systems have been widely adopted in the mainstream. The key benefit to the industry has so far been the reduction of costs, promoting efficiency by eliminating paper passed processes, streamlining end-to-end processes from manufacturing to distribution, and linking to trade financing and payments.

6.5 BLOCKCHAIN-DRIVEN ASSETS – CRYPTOASSETS AND ITS CLASSIFICATION

A Cryptoasset (Ankenbrand et al., 2020) is defined as "a digital representation of value, which a central bank or financial institution may issue, or any private entity or a decentralized ledger (blockchain), which is secured and transacted using cryptographic means." Cryptoassets are emerged as an alternative to legally tendered money. It can also be physically represented as paper printouts or an engraved metal object. Market participants and consumers stand to benefit significantly from these technologies. Financing small and medium-sized enterprises (SMEs) can be more accessible, less costly, and more inclusive with cryptoasset issuances since they streamline capital-raising processes and increase competition with existing financial intermediaries. Through payment tokens, payments can become cheaper, faster, and more efficient, especially on a cross-border basis, as fewer intermediaries are needed. These assets also work on the principle of decentralization, where the assets are not created with the control of one party. They are distributed and backed by a governance protocol. The Bank of International Settlements (BIS) has an opinion that cryptoassets (Basel, 2021) may be used in a legal or nonlegal capacity to exchange between natural and legal persons but not a legal tender. They can be transferred, stored, or traded electronically and used as an alternative to legal money. The consequence of this is that nobody is obliged to accept payments in cryptoassets (except in the case of regulated assets or CBDCs) in any form of physical or digital space (Figure 6.3). The market has been rapidly developing around two broad categories of cryptoassets.

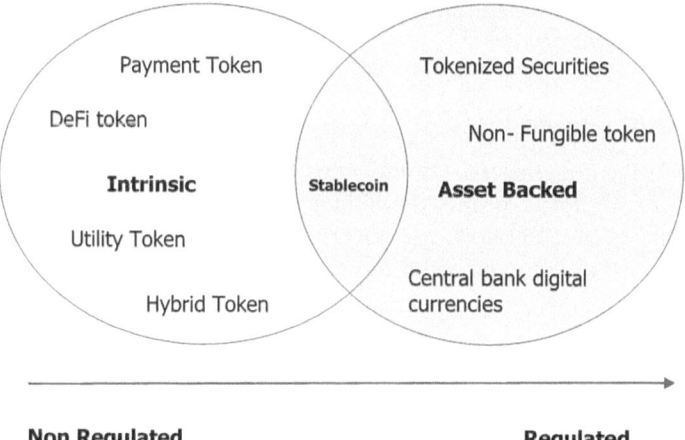

Figure 6.3 Classification of cryptoassets.

6.5.1 Intrinsic cryptoassets

This token does not refer to an existing physical or digital asset but rather to the asset itself. It is stored on a blockchain-based distributed ledger. The asset could be a recognizable financial asset, such as bonds, equity, or bank deposits owed to their issuer. Among native-asset tokens, utility tokens are experiencing rapid growth and can be further divided into three categories.

6.5.1.1 Payment tokens (cryptocurrencies)

A payment token can be regarded as a financial instrument that may also be regarded as a speculative asset. Retail consumers, who are comfortable with market and volatility risks, use exchange tokens for certain niche payment types. Cryptocurrencies (Schletz et al., 2020), according to the Bank for International Settlements, have three components. Primarily – A protocol defines how transactions are carried out (the rules). In addition, there is a ledger that records transactions and lastly a distributed network of participant updates, stores, and reads the ledger using predefined smart contracts.

6.5.1.2 DeFi token

The emergence of cryptocurrencies and their inherent features, such as machine-based, peer-to-peer and smart contracts, has led to the rise of a decentralized digital economy, also known as DeFi. The DeFi network has led to the creation of newer business models, such as decentralized exchanges and automated liquidity systems or staking that operate outside the regulatory framework.

6.5.1.3 Utility token

A utility token facilitates nonmonetary transactions, governance, and access. A utility token grants the holder proportional rights in the governance policies of a network for governing accessing services that utilize machine-based or computational transactions.

6.5.1.4 Hybrid token

The token provides both utility and payment functions (such as Ether on the Ethereum network). It is used as a means of payment to buy other tokens in ICOs. It is also used to pay the network fees for transactions or "gas." Its derivative tokens are used in governance in many blockchain protocols.

6.5.1.5 Asset-backed tokens

6.5.1.5.1 Tokenized Securities

Tokenized securities (BIS, 2021) are investment assets that are only available through blockchain or DLT, including proof of ownership. Securities tokens are digital, liquid contracts for fractions of any existing asset with value, such as a large real estate property. Retail investors now have access to high-value asset classes due to the democratization of these assets.

6.5.1.5.2 Central Bank Digital Currencies (Regulated)

CBDC is a payment settlement token which is the digital equivalent of physical notes and coins backed by central banks as another form of money alongside bank reserves and cash. CBDCs are widely regarded as a risk-free alternative to cryptocurrencies like Bitcoin and stablecoins (Chevet, 2018).

6.5.1.5.3 Non Fungible Tokens (NFTS)

The NFT (Arner et al., 2020) is a cryptoasset representing ownership of real world or virtual items like music, art, in-game items, properties, and goods. It is commonly used for buying and selling such items over a marketplace with or without cryptocurrency. NFTs establish a public record of ownership and ownership verification using blockchain technology.

6.5.1.6 Stablecoins

Stablecoin tokens (Lyons & Viswanath, 2020) are primarily used for payment settlement and to ensure a stable exchange rate. A token's underlying asset can range from a single fiat currency (pegged 1:1) to those that rely on algorithmic mechanisms to maintain price stability. Initially, stablecoin was introduced as a hedge against volatility in cryptocurrency values. Its primary purpose is to act as an intermediary between fiat currencies and cryptocurrencies. Many jurisdictions have introduced regulations (Malloy & Lowe, 2021) such as financial institution mandates, to protect customers and ensure accountability to issuers. Institutions and regulators are examining the potential application of stablecoins in mainstream markets outside of the crypto economy. There are several ways in which it could be used, including improving the efficiency of domestic and cross-border payments or as an investment asset. Due to this, this token type's risk profiles and potential appeal differ significantly. Stablecoins remain subject to ongoing debates about governance and regulations beyond the crypto economy. Stablecoin arrangements perform vital functions, including creation and redemption, transfer between parties, and storage by users, most of

Table 6.2 Types of stablecoin

Type	Definition	Examples
Fiat collateralized	A stablecoin that uses fiat currency and is backed at a 1:1 ratio, means that one stablecoin is backed by 1 unit of fiat currency	USDT, USD Coin (USDC)
Commodity collateralized	A stablecoin backed by a commodity is an exchangeable asset. Gold is often used as collateral. Stablecoins are also supported by oil, real estate, and precious metals.	Digix Gold (DGX), SwissRealCoin (SRC)
Crypto collateralized	Stablecoins are backed by other cryptocurrencies. Since all transactions are conducted through blockchain technology, crypto-backed stablecoins are more decentralized than fiat-backed stablecoins.	Dai, Jarvis FIAT (jFIATs)
Algorithmic	The supply of stablecoins is controlled by an algorithm governed by the coin's operating mechanism. This is the seignorage shares model. In this model, to reduce the price back to normal, new stablecoins are also created; the circulating supply of a coin is reduced if the price is too low. Therefore, stablecoin prices would remain stable as demand and supply determine their prices.	Ampleforth (AMPL), Terra

which involve different activities (Money, 2022) (Table 6.2). Broadly, there are four types of stablecoin.

The stablecoin issuers and entities that plan projects see them as more than just a digital asset used by consumers to pay for goods and services, corporations to complete supply chain payments, and international remittances. There is a high chance that the amount to which stablecoins will be used for these purposes will be determined by the degree of confidence that users have in the stablecoin issuer, including their confidence in its ability to maintain a constant value and allows redemption. Despite this, the transition to more widespread use of stablecoins could occur quickly due to network effects or by leveraging existing platforms and user bases (Ferreira, 2021).

Stablecoins such as Tether (USDT), USD Coin, Binance USD, and Dai (DAI) are already essential to the cryptoasset ecosystem. Their structural stability is expected to make them more useful than cryptocurrencies, which are transparent, auditable, rapid, and cost-effective. In addition to daily payments and international transactions, banks and industries can use them to convert to fiat currencies or to make payments. Traditional settlements with expensive legacy systems can sometimes take days, and can be reversed, posing counterparty risks (Bechtel et al., 2020). In contrast to fiat currencies, stablecoins move frictionlessly through virtual payment systems.

Moreover, apps can be built to interact with stablecoins, enabling a range of capabilities, from simple payment automation to complex business logic. Based on these, a number of business models and use cases have been developed using blockchain technology. They serve as a backbone of emerging decentralized finance (or DeFi) and contribute significantly to the development of crypto trading. Compared to traditional cryptocurrencies, they have relative stability. Consequently, they offer complete access to digital currencies such as "digital cash" and daily transactions. This bridges traditional financial markets with emerging opportunities offered by cryptocurrency technology (Gordon & John, 2022).

Case Study: USDT (a stable coin pegged against the US dollar)

Tether is a fiat-collateralized stablecoin backed with the US dollar. The Tether cryptocurrency is the third largest in circulation after Bitcoin and Ethereum. Bitcoin can be purchased using Tether, one of the most popular options. According to recent speculation, the Tethers issued without underlying reserves are driving the Bitcoin price rise. Tether is a fiat-collateralized stablecoin, and simplicity is the best thing about stablecoins. As long as the country's economy stays stable, the value of a pegged coin shouldn't fluctuate much. This isn't always the case despite stablecoin issuers claiming 100% fiat backing. Stablecoin issuers might store cash reserves in other assets, such as corporate bonds, secured loans, or investments. In recent years, Tether (USDT), the most popular USD-backed stablecoin, has attracted controversy as its 1:1 stablecoin-to-fiat ratio has been questioned. CFTC investigation (Ehrsam, 2016) found that from 2016 to 2019, Tether falsely claimed to hold an equivalent amount of fiat currency for every USDT. The CFTC fined Tether $41M in October 2021. Only 10% of Tether's reserves were in cash and bank deposits, far from the 1:1 ratio reported by its attestation report.

Stablecoins are used to settle automated financial products through smart contracts which are self-executing software based on predefined conditions. Global stablecoin proposals will allow new forms of online exchange due to their 24/7 accessibility, borderless nature, fractionalization, and integration with nonfinancial services. In this way, they aim to challenge existing e-commerce payment methods such as credit cards and bank payment electronic wallets. Stablecoins provide a long-term perspective on a financial ecosystem's inefficiencies and can complement the current payment and investment system. To reap the benefits of stablecoins, regulators must take a long-term approach. A stablecoin ecosystem should be supported by regulations that do not stifle innovation and encourage competition.

6.6 BLOCKCHAIN AS THE FUTURE OF FINANCE

DeFi is redefining the future of Finance (DeFi). The underlying infrastructure that powers financial applications is undergoing a significant shift, and it is changing how we think about permissions, controls, transparency, and risks. DeFi combines blockchain technologies, digital assets, and financial services (Decentralized, 2013). A DeFi app is a financial application that does not have a central counterparty. It means that, in practice, one is not interacting with an institution (such as a bank) to access these financial applications. Instead, users interact directly with the programs (such as smart contracts) on top of the protocol (Auer & Claessens, 2020). DeFi apps share standard features, including integration with blockchain technology as the core ledger, open source, and transparent by default which can be audited and forked by anyone for specific purposes, programmable, and interoperable among similar blockchain-based platforms (majorly Ethereum as blockchain). DeFi measures profitability using total value locked (TVL). Creating a stable, secure network involves staking (locking) digital assets for a set period. Using the over/undervalue tracked by DeFi Pulse, TVL ratios are calculated by multiplying supply in circulation by price to get the market cap, and then dividing it by supply. DeFi's ecosystem is built on public blockchains at the low layer (settlement), digital assets (tokens) at the asset layer, and dApps (Igor & Igor, 2021) (Figure 6.4). It has four layers:

Layer 1: Settlement layer – Blockchain as a guarantor and foundational infrastructure of trust, security, and finality.
Layer 2: Asset Layer – Ability to facilitate different cryptoassets to participate in the activities performed over the blockchain. This includes smart contracts driving complex functions for lending, borrowing and trading assets, and payments.
Layer 3: Application layer – Sophisticated financial services built over asset layer to facilitate digital applications called as dApps
Layer 4: Aggregation layer – Direct access to aggregated offerings through a friendly user interface, allowing such services as storing and sending money, investing and exchanging cryptoassets, and borrowing against the assets.

DeFi token holders vote on issues on the network by participating in self-governance. Crypto mining is one way to prevent double spending using computational power. DeFi has scalability issues and usage fees. There are a number of factors holding back mainstream adoption of ESG (environmental, social and corporate governance) framework, including slow speeds, high costs, and the use of electricity, which is the opposite of ESG in financial services. The Ethereum network is utilized by most DeFi projects

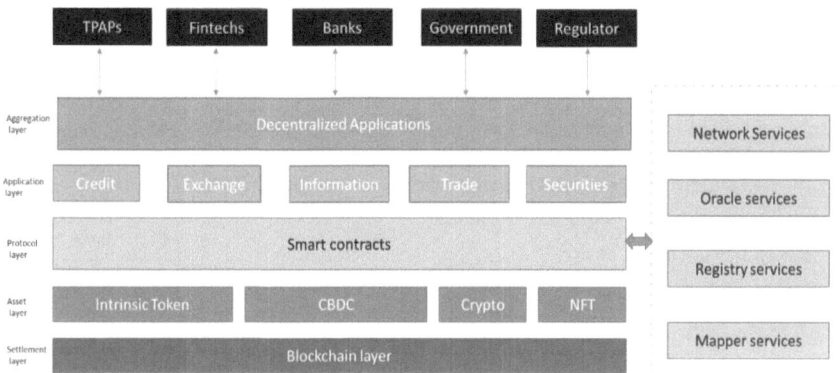

Figure 6.4 The DeFi ecosystem.

thanks to smart contracts, which are essential for DeFi applications. As a result, Ethereum's Solidity programming language offers developers a wide range of options for creating smart contracts. On a blockchain, a smart contract executes a set of predefined instructions. Vending machine is a famous example of a smart contract. A smart contract is written in solidity language (Insights, 2022) as illustrated below:

Illustration: Smart contract for vending machine having Lassi taking ether as payment tokens [<Ref: Ethereum Developer Documents>]

```solidity
pragma solidity;

contract VendingMac {

  // Declare state variables of the contract for user's
Vending Machine for Lassi
  address public owner;
  mapping (address => uint) public lassiBalances;

  // When 'VendingMachine' contract is deployed:
  // 1. set the address as the User being owner of the
       smart contract
  // 2. set the smart contract's lassi units to 50
  constructor() {
    owner = msg.sender;
    lassiBalances[address(this)] = 50;
  }
```

```
// Allow the owner to increase the smart contract's lassi
balance
  function refill(uint amount) public {
    require(msg.sender == owner, "Only the owner can
refill.");
    lassiBalances[address(this)] += amount;
  }

  // Allow anyone to purchase lassis from Ashutosh Vending
machine
  function purchase(uint amount) public payable {
    require(msg.value >= amount * 1 ether, "You must pay at
least 1 ETH per lassi");
    require(lassiBalances[address(this)] >= amount, "Not
enough lassis in stock to complete this purchase");
    lassiBalances[address(this)] -= amount;
    lassigBalances[msg.sender] += amount;
  }
}
```

Tokens form the basis for the DeFi ecosystem, as smart contracts are themselves tokens. Smart contracts can be either simple or complex. The blockchain can be used to build corporate entities such as decentralized autonomous organizations (DAOs) with multiple smart contracts, including a digital mutual fund. Since all events are logged and immutable on the blockchain, even the most complex DAO can have a single chain of custody, while all events are logged and immutable. DAOs are what make up DeFi. Emergence of metaverse is an important milestone to make blockchain as a foundational infrastructure of future.

METAVERSE – PRACTICAL APPLICATION OF WEB 3.0 AND CRYPTOASSET

Our physical and digital lives are seamlessly combined in the metaverse, which creates a unified, virtual community in which people from all over the globe can work, play, relax, conduct business, and socialize. Currently, people own $41 billion worth of NFTs also around 200 (Ball, 2021) strategic partnerships happened in sandbox (a metaverse platform), including the launch of Warner group, Justin Bieber concert. Metaverse offers options to transact $54 billion (Intelligence, 2021) on virtual goods, which is almost double the amount spent on music, socialize with 60 billion (Opportunities, 2022) messages sent daily

(Continued)

on Roblox, and create GDP worth $651 million through Second Life, where $80 million goes to US creators.

Social interactions can be deepened and extended in many virtual worlds that are taking shape to support this vision of the metaverse. This vision is being supported by new technologies. The user experience has improved with augmented reality (AR) and virtual reality (VR) headsets. Digital currencies and NFTs can now be generated and used on blockchain. With the new methods of transacting and owning digital goods, creators are able to monetize their activities through tokens. In addition to monetization, token holders can also participate in the platform's governance (e.g. vote on decisions) as well as exchange value. Combined with the potential for interoperability, such a democratic ownership economy could unlock immense economic opportunities, whereby digital goods and services will no longer be limited to a single gaming platform or brand. Increasingly, immersive virtual experiences are helping people to build communities based on shared values and to express themselves in more authentic ways. COVID-19, meanwhile, has accelerated the digitization of our lives and normalized continuous and multipurpose online engagement and communication. A combination of technological, social, and economic factors is responsible for the increase in interest in the metaverse.

The term and concept of metanomics (The Metaverse, n.d.), however, is not a new one. The origin of the movement can be traced back to 2007, when Rob Bloomfield, a Cornell professor, ran a course on the topic in second life. Today many of the themes Bloomfield covered are still relevant, including drawing parallels between physical and digital real estate. Today, however, a key difference is the advent of Web 3.0, which has given rise to an ownership economy. It is possible to purchase an original piece of art, tokenized as a digital asset, as a way of personalizing your virtual home. The virtual land that is used for building a home can even be owned by the owner. A number of Ethereum-based platforms, for example Decentraland, are already selling virtual plots (Abadi, 2018) of land that people can develop. There is an increasing market for virtual real estate. It is predicted that the average price of a parcel of land will double in a period of 6 months between 2021 and 2022. Across the four main Web 3.0 metaverses, it jumped from $6,000 in June to $12,000 by December this year.

Metaverse has the potential to massively expand access to the market for consumers from emerging and frontier economies. Internet access has already opened up a world of goods and services previously unavailable. A clear governance framework will be essential to these developments. There

(Continued)

is the possibility of massive scalability from a corporate perspective. An international retailer might build a global hub in the metaverse instead of opening stores in every city to serve millions of customers. It is now possible, for example, for workers in low-income countries to obtain jobs in western companies without having to emigrate. VR worlds have the potential to extend educational opportunities, as they can be utilized for training at a low cost. The metaverse is poised to accelerate the shift from cash to crypto in areas like gaming, sports, and trading.

Using cryptocurrency as a payment method can circumvent conventional card and mobile payment methods. For asset managers, sovereign wealth funds, and institutional investors, DeFi can increase flexibility and cut costs by managing partnerships, custodianships, and vendor relationships. Companies from all industries are now using blockchain to manage their financial operations, creating tokens and native financial tokens closely linked to their businesses. Currently, Web 3.0 is majorly built on Ethereum blockchain, so most smart contracts, dApps, and DeFi are built on them. They are not interoperable with other blockchains, which are not built on Ethereum as standard. The major downside is network congestion, which raises gas prices and transaction costs. This can potentially make Web 3.0 very expensive. Several advantages come along with DeFi, but there is a trilemma in that improving speed, and scalability causes a reduction in another characteristic, like the level of decentralization or security.

6.7 BLOCKCHAIN-ASSOCIATED RISKS

The existence of decentralized databases has been recognized by computer scientists as far back as the early 1980s. More recently, a variation to this trilemma (Im, 2018), known as the "Blockchain Trilemma," was termed by Ethereum creator Vitalik Buterin, where he proposed that a blockchain can possibly support two of three pillars at one time: decentralization, security, or scalability.

Network security refers to the likelihood of an attack or compromise. Decentralized networks are prone to 51% attacks, which are circumstances in which a single person or entity controls more than half of the processing power (the hash rate) of a network. A successful attack could allow the attacker to control the network or, more specifically, perform double spending, jacking new tokens or reverse transactions. The level of decentralization refers to how much control any one person, group, or entity has over a system or network. A voting mechanism or solving a complex problem achieves consensus in a decentralized network. No one entity has control over the data in a decentralized network. Open, decentralized networks are open to anyone as long as they follow the rules and protocol intermediaries. As a result of the need for a more substantial consensus, higher decentralization lowers the network's throughput or the speed at which information can travel. A decentralized network is the opposite of a centralized network controlled by one intermediary. As there does not need to be a consensus, this allows for incredible speed and throughput, but at the cost of trusting a single agent. The scalability of a network has to do with how well it can handle growth, such as growing the number of users and how many transactions it can process in a limited amount of time.

6.8 CONCLUSION

Web 3.0 applications rely on blockchain, the decentralizing backbone. Blockchain technology enables multiple parties to work together with shared assumptions and data without trusting each other. The information could be anything, such as the location and destination of an item in a supply chain or account balances tied to tokens. Smart contracts power it. On top of a blockchain, smart contracts can create and alter arbitrary data or tokens. With this software, rules can be efficiently encoded for any transaction, and even specialized assets can be created. A smart contract could enumerate and algorithmically enforce any traditional business contract. Data stewardship, gaming, and supply chain all benefit from smart contracts. Blockchain protocols are isolated from the world outside of their ledger. Oracles are used to solve this problem. Oracle is any source of external data for reporting information in the context of smart contract platforms.

Since many cryptocurrencies are volatile, users who want to use Decentralized applications but are not comfortable with investing in a volatile asset like cryptocurrency will have difficulty doing so. Stablecoins, a class of Cryptoassets, have been created to address this problem. They offer investors the consistency they desire when participating in several Web 3.0 applications by maintaining price parity with a target asset, such as USD, legal tender, gold, or any commodity. To establish a DAO, blockchain has laid the foundation with smart contracts that identify who may perform what actions and upgrade using governance tokens. The tokens give owners some influence on future outcomes of the platform, making it more democratic. Many service providers have already begun experimenting with blockchain-based products and services. The decentralized finance industry may benefit from traditional banks' expertise in acquiring customers, offering integrated services, and managing the back office. Their services could include custody, tokenization, crypto, and decentralized financial infrastructure. Crypto companies might also get tailored banking services or capital market trading capabilities. In banking, blockchain could be used for settlement systems, payment transfers, CBDC, etc. It could also transform risk and compliance from a cost center to a strategic partner. Blockchain technology can be used for risk management and compliance by merchants and payment companies. Various forms of payment and trading may be enabled by the technology, such as super apps, platform as a service, etc. All the major global forums are discussing the framework for a magnificent new ecosystem based on Web 3.0. This is not just a reconstruction of existing laws of the land, but one built from the ground up. Blockchain could be transformative and can met. Through its transparency, security, and trust in actions, blockchain may address organizational problems. Web 3.0 or DeFi may realize its potential, and companies who refuse to adapt may be lost in the coming decade. The blockchain allows a single view of information based on agreed-upon values, and it permits self-executing actions based on those values.

REFERENCES

Abadi, J., & Brunnermeier, M. (2018). *Blockchain Economics (No, w., 25407)*. National Bureau of Economic Research.

Adrian, T., & Mancini-Griffoli, T. (2021). The rise of digital money. *Annual Review of Financial Economics*, 13(1), 57–77. https://doi.org/10.1146/annurev-financial-101620-063859.

Allende, R., et al. *Cross-Border Payments with Blockchain*. Inter-American Development Bank. https://publications.iadb.org/publications/english/document/Cross-Border-Payments-with-Blockchain.pdf.

Alvarez, F., & Argente, D. (2020). *Central Bank Digital Currencies: Foundational Principles and Core Features*. Bank of International Settlements.

Ankenbrand, T., Bieri, D., Cortivo, R., Hoehener, J., & Hardjono, T. (2020). Proposal for a comprehensive (crypto) asset taxonomy. In *Cryptographic Valley Conference on Blockchain Technology (CVCBT), 2020* (pp. 16–26). IEEE Publications.

Antonopoulos, A. M., & Wood, G. (2018). *Mastering Ethereum: Building Smart Contracts and Dapps*. O'Reilly Media.

Arner, D. W., Auer, R., & Frost, J. (2020). Stablecoins: Risks, potential and regulation. *SSRN Electronic Journal*. https://doi.org/10.2139/ssrn.3979495.

Auer, R., & Claessens, S. (2020). *Cryptocurrency Market Reactions to Regulatory News. Globalization Institute Working Paper 381*, Federal Reserve Bank of Dallas.

Aysan, A.F., Bergigui, F., Disli, M. (2021). Using blockchain-enabled solutions as SDG accelerators in the international development space. *Sustainability*, 13, 4025. https://doi.org/10.3390/su13074025.

Ball, M. (2021). *Payments, Payment Rails, and Blockchains, and the Metaverse*. Matthewball.vc. Retrieved February 20, 2022. https://www.matthewball.vc/all/metaversepayments.

Banerjee, M., Lee, J., & Choo, K. R. (2018). A blockchain future for internet of things security: A position paper. *Digital Communications and Networks*, 4(3), 149–160. https://doi.org/10.1016/j.dcan.2017.10.006.

Basel Committee. (2021). *Prudential Treatment of Cryptoasset Exposures*. Basel Committee on Banking Supervision, Bis Consultative.

Bechtel, A., Ferreira, A., Gross, J., & Sandner, P. G. (2020). *The Future of Payments in a DLT-Based European Economy: A Roadmap*. http://doi.org/10.2139/ssrn.3751204.

Bis, H. S. S. (2021). Chapter III. Mode of access. Electronic resource. In *CBDCs: An Opportunity for the Monetary System, Annual Economic Report* (Date of access 20. 10. 2021). https://www.bis.org/publ/arpdf/ar2021e3.htm.

Blockchain Market Size, Share and Global Industry Trend Forecast till 2025. (n.d.). Retrieved February 20, 2022. https://fortunebusinessinsights.com/enquiry/request-sample-pdf/blockchain-market-100072.

Chang, S. E., Luo, H. L., & Chen, Y. (2020). Blockchain-enabled trade finance innovation: A potential paradigm shift on using letter of credit. *Sustainability*, 12(1), 188. https://doi.org/10.3390/su12010188.

Chen, Y., & Bellavitis, C. (2020). Blockchain disruption and decentralized finance: The rise of decentralized business models. *Journal of Business Venturing Insights*, 13, e00151. https://doi.org/10.1016/j.jbvi.2019.e00151.

Chevet, S. (2018). Blockchain technology and non-fungible tokens: Reshaping value chains in creative industries. Available at SSRN 3212662.

Chowdhury, M. J. M., Ferdous, M. S., Biswas, K., Chowdhury, N., Kayes, A. S. M., Alazab, M., & Watters, P. (2019). A comparative analysis of distributed ledger technology platforms. *IEEE Access*, 7–167943. https://doi.org/10.1109/ACCESS.2019.2953729, PubMed: 167930.

Decentralized Finance (DeFi)—Risks and Opportunities for the Insurance Industry. International Insurance Society. (n.d., September 13). www.internationalinsurance.org. Retrieved February 20, 2022. https://www.internationalinsurance.org/Insights_decentralized_finance.

Ehrsam, F. (August 1, 2016). Blockchain tokens and the dawn of the decentralized business model, *Coinbase Blog*. https://blog.coinbase.com/app-coins-andthe-dawn-of-the-decentralized-business-model-8b8c951e734f.

Ferreira, A. (2021). The curious case of Stablecoins—Balancing risks and rewards? *Journal of International Economic Law*, 24(4), 755–778. https://doi.org/10.1093/jiel/jgab036.

Gonzalez, L. (2019). Blockchain, herding and trust in peer-to-peer lending. *Managerial Finance*, 46(6), 815–831. https://doi.org/10.1108/MF-09-2018-0423.

Houben, R., & Snyers, A. (2020). *Crypto-Assets. Key Developments, Regulatory Concerns and Responses. Policy Department for Economic, Scientific and Quality of Life Policies Directorate-General for Internal Policies*. PE, 648, 13.

Igor, M., & Igor, S. (2021, July 14). *Defi in a Nutshell*. ThePaypers. Retrieved February 20, 2022. https://thepaypers.com/expert-opinion/defi-in-a-nutshell-1250375.

Iivari, N., & Sharma, S. (2020). Leena Ventä-Olkkonen, Digital transformation of everyday life – How COVID-19 pandemic transformed the basic education of the young generation and why information management research should care?, *International Journal of Information Management*, 55, https://doi.org/10.1016/j.ijinfomgt.2020.102183.

Im, D. K. D. (2018). *The Blockchain Trilemma.*

Insights, L. (2022, January 27). *Warner Music Partners with Sandbox Blockchain Game for Music Events.* Ledger Insights—Enterprise Blockchain. Retrieved February 20, 2022. https://www.ledgerinsights.com/warner-music-partners-with-sandbox-blockchain-game-for-music-events/.

Intelligence, B. (2021, November 5). Into the metaverse: Developments, investments & experiences in the metaverse on apple podcasts. *Apple Podcasts.* Retrieved February 20, 2022. https://podcasts.apple.com/us/podcast/developments-investments-experiences-in-the-metaverse/id1593908027?i=1000540906629.

Kulk, E., & Plompen, P. (2021). Demystifying programmable money: How the next generation of payment solutions can be built with existing infrastructure. *Journal of Payments Strategy & Systems*, 15(4), 445–454.

Liao, G. Y., & Caramichael, J. (2022). *Stablecoins: Growth Potential and Impact on Banking.* Board of Governors of the Federal Reserve System International Finance Discussion Papers.

Lyons, R. K., & Viswanath-Natraj, G. (2020). *What Keeps Stablecoins Stable?* NBER Working Papers.

Malloy, M., & Lowe, D. (2021). Global Stablecoins: Monetary Policy Implementation Considerations from the US Perspective. FEDS Working Papers.

Momtaz, P.P. (2020). Initial coin offerings. *PLoS One*, 15(5), e0233018. https://doi.org/10.1371/journal.pone.0233018.

Money and Payments: The U. S. Dollar in the Age of Digital Transformation. (2022, January). Retrieved February 20, 2022. https://www.federalreserve.gov/publications/files/money-and-payments-20220120.pdf.

Nakamoto, S. (2008). Bitcoin: A peer-to-peer electronic cash system. *Decentralized Business Review*, 21260.

Opportunities in the Metaverse—jpmorgan.com. (n.d.). Retrieved February 20, 2022. https://www.jpmorgan.com/content/dam/jpm/treasury-services/documents/opportunities-in-the-metaverse.pdf.

Rauchs, M., Blandin, A., & Dek, A. (2020). *Cambridge Bitcoin Electricity Consumption Index (CBECI).* https://cbeci.org/mining_map. Accessed May 30 2020.

Reiff, N. (2021). *What's the Environmental Impact of Cryptocurrency.*

Rudman, R., & Bruwer, R. (2016). Defining Web 3.0: opportunities and challenges. *Electronic Library*, 34(1), 132–154. https://doi.org/10.1108/EL-08-2014-0140.

Schletz, M., Nassiry, D., & Lee, M. K. (2020). *Blockchain and Tokenized Securities: The Potential for Green Finance.*

The 2021 Metaverse Real Estate Report. (n.d.). Retrieved February 20, 2022. https://republicrealm.docsend.com/view/9fdnrtcsxh9u9wn2.

Wild, J., Arnold, M., & Stafford, P. (2015). Technology: Banks seek the key to blockchain. *Financial Times.*

Zarrin, J., Wen Phang, H., Babu Saheer, L., & Zarrin, B. (2021). Blockchain for decentralization of internet: Prospects, trends, and challenges. *Cluster Computing*, 24(4), 2841–2866.

Chapter 7

Applications of blockchain technology for sustainable education

Franklin John and Nikhil V. Chandran
Kerala University of Digital Sciences, Innovation and Technology

CONTENTS

7.1 INTRODUCTION

Blockchain technology came into existence with the introduction of Bitcoin. Unlike other digital currencies before, Bitcoin was able to gain popularity and adoption. Anyone with an internet connection can be part of the network. Many factors led to the popularity of the Bitcoin network, like its decentralised architecture, immutability, cryptographic security, pseudo-anonymity, traceability and double-spending protection. Some of these properties like decentralisation and double-spending protection were absent in the predecessor currencies of Bitcoin. All these were made possible by the technology on which Bitcoin was implemented – blockchain.

Blockchain is a type of distributed ledger. The distributed ledger acts as a record-keeping ledger system to store all the transactions occurring in the network. Unlike a centralised system where all the records are kept at a single point, the distributed ledger system uses a computer network to store the ledger. Each participating node in the network will store a copy of the ledger and will update it simultaneously. Transactions coming to the network are grouped and stored as a block, and these blocks are linked in chronological order; this is done by keeping the previous block's hash value in the newly created block. Thus, it becomes practically impossible to alter the data written in a blockchain. Even if some node in the network manages to tamper with the data in the blockchain, due to its decentralised nature, the blockchain network will not accept the altered blockchain.

As Bitcoin gained popularity, efforts were made to improve the underlying blockchain technology. Even today, only an average of seven transactions can be processed by the Bitcoin network; this is a drawback because other transaction processing systems like Visa claim to process 65,000 transactions/second (Visa Fact Sheet, 2018). The operations that can be done on the Bitcoin network are restricted to Bitcoin transactions. The scripting functionalities are also limited. In order to harness the potential of this globally available decentralised system, alternatives were introduced. The first one among them was the Ethereum blockchain. Ethereum blockchain introduced the concept of smart contracts to the blockchain space, thus starting a new era of blockchains called programmable blockchains.

Ethereum supports deploying smart contracts to its blockchain network. They developed a new programming language called Solidity to write smart contracts; this opened a new array of possibilities, resulting in numerous decentralised applications. Decentralised applications are applications that work with the help of blockchains. Decentralised applications, often called decentralized application (DApps), are developed for various domains, including finance, gaming, media, energy and governance. These applications focus on leveraging the inherent properties of blockchain, thus bridging the gap in the adoption of decentralisation in day-to-day use cases. The increasing interest in DApps and programmable blockchains gave birth to more blockchains that try to overcome the drawbacks of existing blockchains. Blockchains like Polkadot, Cardano, Solana, Avalanche, Algorand are gaining traction by introducing innovative ideas to enable DApp development with focus on more decentralisation, interoperability and throughput. Layer 2 frameworks like a lightning network in Bitcoin, Polygon network in Ethereum and techniques like off-chain transactions, rollups like optimistic rollups and zero-knowledge rollups plays an important role in improving the functionalities of blockchain.

Blockchains or distributed ledger technologies (DLTs), in general, can be classified into two broad categories – permissionless and permissioned. Early blockchains like Bitcoin and Ethereum are considered permissionless public blockchains. They are also permissionless. Anyone with internet access and the right equipment can join these blockchain networks and do transactions. The codebase for such networks will be open source, and anyone can suggest updates and improvements to it. The participants in the network need not know each other and are anonymous to a certain degree. Most permissionless blockchains have an associated cryptocurrency to regulate and govern their activities. The data written in the blockchain in such networks will be available to the public, and anyone can check and verify that information; this also raises concerns about data privacy. Not everyone doing transactions on public blockchains may want the details of their transactions to be made available for the public to see. Industries wanted an alternative for the public blockchains (Linux Foundation's Hyperledger Project Announces 30 Founding Members and Code Proposals to Advance Blockchain Technology – Hyperledger Foundation, 2017); this resulted in the development of permissioned blockchains.

As opposed to permissionless blockchains, permissioned blockchains impose restrictions on who can join the network, write to the blockchain and read from the blockchain. There will be an authority controlling the activities in the network, and the network participants will be of known identities. Since only restricted parties can join the network, regulating and governing the activities will be easy. Even though many permissioned blockchain systems support cryptocurrency tokens, they do not need one necessarily. Technologies like Hyperledger Fabric, Hyperledger Sawtooth and R3 Corda are examples of a permissioned blockchain network. These networks provide the functionalities to control the participants joining the network, restrict read and write permissions to the participants, manage identities and integrate with legacy systems and DApp deployment. These permissioned distributed ledgers can be combined with permissionless blockchains to develop hybrid systems that leverage the features of both these systems.

Blockchain offers a new perspective on how we perceive current technologies. How decentralisation, immutability and traceability help in the current technological society. Blockchain has already started disrupting the finance, banking and supply chain industries. More industries are experimenting with how blockchain adoption can help them.

In this chapter, we try to explore the possibilities of how adopting blockchain in the field of education will help meet the sustainable development goals proposed by the United Nations (THE 17 GOALS|Sustainable Development, 2022). The sustainable development goals envision a set of actions centred on people, the planet and prosperity. The 17 goals aim to eradicate poverty and hunger, provide good health, quality education, clean water and sanitation, affordable and clean energy and remove inequalities in society. These goals are planned to be achieved by the global partnerships between governments, the private sector, civil society, the United Nations system and other actors (Linux Foundation's Hyperledger Project Announces 30 Founding Members and Code Proposals to Advance Blockchain Technology – Hyperledger Foundation, 2017). From a technological point of view, blockchain can help achieve sustainable development goals in various domains, as proposed by the United Nations Conference on Trade and Development (Harnessing blockchain for sustainable development: prospects and challenges|UNCTAD, 2022).

Major contributions of this chapter:

- A comprehensive review of the major literature works in the domain of blockchain for education is provided.
- This chapter gives an overview of the works done to integrate blockchain with the education sector.
- The chapter also discusses case studies to incorporate blockchain in the education sector and how it helps in achieving the sustainable development goals of the United Nations.

The remainder of this chapter is organised as follows. Section 7.2 presents some related works about blockchain and its applications in the field of education. Sections 7.3.1 and 7.3.2 present two case studies for blockchain in education, "CertiChain" a blockchain-powered academic certificate issuance and verification system and "Sign-A-Doc" a blockchain-powered multi-party document signing and verification system, respectively. Then, some future research dimensions are discussed in Section 7.4. Finally, a summary of this chapter is given in Section 7.5.

7.2 LITERATURE REVIEW

Blockchain as an underlying technology was introduced in 2008 when an anonymous individual or a group of individuals known as "Satoshi Nakamoto" published a whitepaper titled "Bitcoin: A Peer to Peer Electronic Cash System." The paper proposed a new kind of financial system which does not depend on a trusted third party to function. The popularity of Bitcoin led to the creation of thousands of alternative currencies such as Litecoin and Dogecoin. With the creation of the Ethereum blockchain came the innovation of smart contract-based decentralised applications. The development of Ethereum led to newer models of tokenisation and decentralised finance. Enterprise blockchain came into existence with the Hyperledger Foundation, which focuses on DLTs. Potential application of blockchain and DLT includes banking, finance, supply chain, government and education. The use of blockchain in the education sector is diverse, and research is ongoing in many related areas.

Blockchain technology has the advantages of decentralisation, immutability and tamper resistance, enabling its use in education. The most important role of blockchain in education is the issuance and storage of educational records. These records can be academic certificates, learning materials or transcripts stored securely on a blockchain. In recent years, there has been an increase in the number of fake certificates and transcripts used illegally. Research has shown that counterfeiting certificates are done in an organised manner by some government and university officials (UntungRahardja and

EkaPurnamaHarahap, 2020). Blockchain-based solutions are proposed for the issuance and tracking of certificates to solve this problem. The idea is that blockchain could help to reduce the time for verification of the certificates and reduce the cost of storage. Fake certificates could be traced easily on the blockchain and smart contracts could eliminate the middle parties (Caldarelli and Ellul, 2021).

Academic certificates were first issued on the Bitcoin blockchain by the Massachusetts Institute of Technology (Kolvenbach et al., 2018). "Blockcerts" developed by MIT is the first blockchain application to store and share digital diplomas and certificates. Blockcerts is blockchain agnostic and can work with Bitcoin, Ethereum or Hyperledger (Ocheja et al., 2019). The University of Singapore has issued certificates on the Ethereum blockchain known as the "Opencerts" program (Koh, 2019). A blockchain research organisation, Kerala Blockchain Academy, has issued certificates on Electro-Optical System (EOS) blockchain known as the CertiChain project. Some other projects were created using the Hyperledger Foundation distributed ledgers for a permissioned network. As mentioned above, the current application of blockchain for academic records is fragmented across multiple blockchain platforms and technologies. The lack of a common standard for creating and issuing academic records hinders the widespread adoption of blockchain in this field. Also, "Oracles," are external third parties known who supply real-world data such as academic records to the blockchain network (Sharples and Domingue, 2016). Since these Oracles are mostly centralised, the probability of them uploading fake and unauthorised data to the blockchain need to be considered while evaluating the trustability of blockchain records.

Tokenisation in education has created a new opportunity for incentivising education. Tokens are a form of asset class deployed onto the blockchain and represent something of value (Chen, 2018). The system can incentivise learners with tokens for reaching milestones or completing assignments (Steiu, 2020). Teachers can incentivise students to obtain additional knowledge, like a new course or certification. Blockchain-based systems can improve learner engagement and make learning more sustainable (Devine, 2015). Decentralised learning focused on students' personal development is made possible through smart contracts (Swan, 2015). One such initiative is BitDegree (Kamišalić et al., 2019), which offers online courses and tutorials in a gamified manner. BitDegree offers tokenised scholarships for completing studies. Educational organisations like Kerala Blockchain Academy provide academic accreditation and certification on the blockchain.

Maintaining educational transcripts and scores through a blockchain platform was introduced by Sony Global Education in 2017 (Guustaaf et al., 2021). By the use of blockchain technology, the examiner will be able to secure and manage their data. This data can be shared across multiple organisations and platforms. An educational stablecoin called Edgecoin provides an open payment system for educational institutions. Edgecoin is currently creating a "Metaverse University" and offering scholarships for courses. On-Demand Education Marketplace (ODEM) is a blockchain-based platform offering students affordable learning opportunities. It allows students to share verified credentials on the blockchain. A blockchain-based education system consisting of Learning Management System (LMS) server, blockchain server, and a client that the students can access was created (Lee et al., 2021). The blockchain contains student face information, behaviour analysis information, access log and evaluation information. The present pandemic situation has necessitated the need for untact education systems to enable efficient online education. Raimundo and Rosário (2021) performed Systematic Bibliometric Literature Review (LRSB) of blockchain applications in higher education area. This study has proved that blockchain is being used for securing knowledge data and sharing student information. Blockchain solutions have increased efficiency and improved privacy in the higher education sector. A decentralised online review system that allows consortium-based onboarding of subject matter experts was proposed (Garg et al., 2021). This eliminates fake reviews and ensures a trusted review system.

The Blocknet project was created to create a modular online course for interdisciplinary blockchain education (Düdder et al., 2021). This project prepared students with interdisciplinary projects as well as taught the application possibilities of blockchain technology. This project brought together relevant disciplines such as supply chain, management, economics, finance and computer science. There are various opportunities for implementing blockchain in education such as for detecting fake or counterfeit certificates (Lutfiani et al., 2021). Blockchain technology has proven to enhance learning methods to make them more engaging and useful. Securing educational data through blockchain is another use case in education. Many more technological advancements in education is expected with the maturation of blockchain. Recently, a combination of blockchain and machine learning in education was proposed (Shah et al., 2021). This system proposes a decentralised database to store students' academic data. The paper states how blockchain can store students' future predicted data through machine learning algorithms. This disruptive combination of blockchain technology and machine learning can be effective in handling student data without security challenges. A blockchain of learning logs (BOLL) system was proposed as a solution to the problem of learning analytics and personalised learning (Ocheja et al., 2022). This system logs student performance across time, student engagement, learning content and quizzes. This work also presented visualisations to support teaching and learning and validated them through interviews.

The role of blockchain in sustainable higher education is being actively explored (Bucea-Manea-ţoniş et al., 2021). The article analyses the use of blockchain technology for improving performance in Higher Education Institutions (HEI). The paper describes the current state of blockchain technology in higher education. It proposes solutions using blockchain technology that HEIs can use to improve their performance. One of the solutions is to include blockchain to devise solutions for a new social and financial ecosystem (Bartoletti et al., 2018). Blockchain technology can be leveraged to tap into the potential of the underutilised social capital of the world. Collaborative learning can be achieved through sharing knowledge across institutions and universities. Gamification can be a way to increase the effectiveness of teaching and learning with the help of blockchain technology. Social media presence and gamification of learning can provide better learner engagement and learning performance (Antonaci et al., 2019). Present-day social media networks that are centralised have the disadvantage of propagating fake information. Blockchain can provide trust and reliability for verifying all news and information spread across social media. With the ever-increasing cost of formal education, students of low- and middle-income families are finding it difficult to get access to the same.

Students form a part of the cultural capital of the world; their education is of the highest importance for the development of the world. Blockchain has provided an avenue for the unbanked population to get into the mainstream by means of wallets. Blockchain-based payment systems do not mandate Know Your Customer forms nor check for credit scores. Similarly, wallets and decentralised education can be used to reduce the cost of education (Jirgensons & Kapenieks, 2018). Tokenisation and gamification of learning along with wallets and collaborative learning through decentralisation can enable a sustainable future for education.

7.2.1 Case study 1: CertiChain – blockchain- powered academic certificate issuance and verification system

CertiChain is a blockchain-powered academic certificate issuance and verification system developed by Kerala Blockchain Academy, an autonomous innovation centre under Kerala University of Digital Sciences, Innovation and Technology, Thiruvananthapuram, Kerala, India. The system is developed to help academic institutions to issue their certificates into blockchain and make information about the certificate accessible by anyone, anytime, anywhere. The goal of the system is to make the certification process transparent, environment-friendly and tamper-proof. The certificates issued through the CertiChain system are supported by strong cryptographic principles of the blockchain. This helps in the verification of the authenticity of certificates issued. Issuing certificates through CertiChain provides an extra layer of security and credibility by using an immutable, decentralised blockchain that can benefit graduates and employers.

CertiChain system works as follows:

The certificate issuing process starts with the registration of the certificate issuing authority. Once signed in they have the options to set certificate templates, manage courses, assign certificates to courses, add users, upload certificates, validate, issue and display certificates.

The certificate template can be set by adding attributes and respective types for each certificate. The attributes may include certificate number, name, issuer name, date of issuance and other information to be included in the certificate. Once the certificate template is created, the issuer can create a course using the course management module. Through this module, the issuer can add courses and set the certificate template for that particular course. This module also offers the option to add and modify batches for the courses. After assigning a certificate template to the course, the issuer can add the details of candidates eligible for certification. This can be done in bulk by uploading a .csv file. The issuer has the option to verify these details through the system. Once verified the issuer can proceed with certificate issuance.

The certificate issuance process comprises three steps.

Step 1: Certificate generation: In this step, a pdf of the certificate is generated and stored in a database.

Step 2: Adding certificate details to blockchain: Once the certificate is generated the details including the certificate number and issuance date are added to the blockchain and the corresponding transaction ID is received.

Step 3: Email notification: The issuer can opt for the feature of email notifications to the candidates whose certificates are issued.

The workflow of CertiChain is illustrated in Figure 7.1. Once the certificates are issued, anyone can verify the authenticity of the certificates by visiting the verification portal, as seen in Figure 7.2. The verification can be done by entering the certificate number in the portal or by scanning the QR code embedded in the certificate. The certification portal will show if the certificate is valid or not. On successful verification, the verifier can view the certificate along with the blockchain transaction ID. The blockchain transaction ID can be used to view the transaction details in a blockchain explorer, as seen in Figure 7.3, thus confirming the authenticity of the certificate.

Key features of the system are as follows:

- **Fully Customisable:** The certificate details are fully customisable and the issuer can add the templates and details according to their requirements.
- **Environment-Friendly:** The paperless certificate issuance and verification enables an environment-friendly certificate issuance process.
- **Instant Verification:** The verification of the certificates is made easy by adding a QR code specific to each certificate. Scanning of the QR code will redirect to the verification portal and thus check the authenticity of certificates.

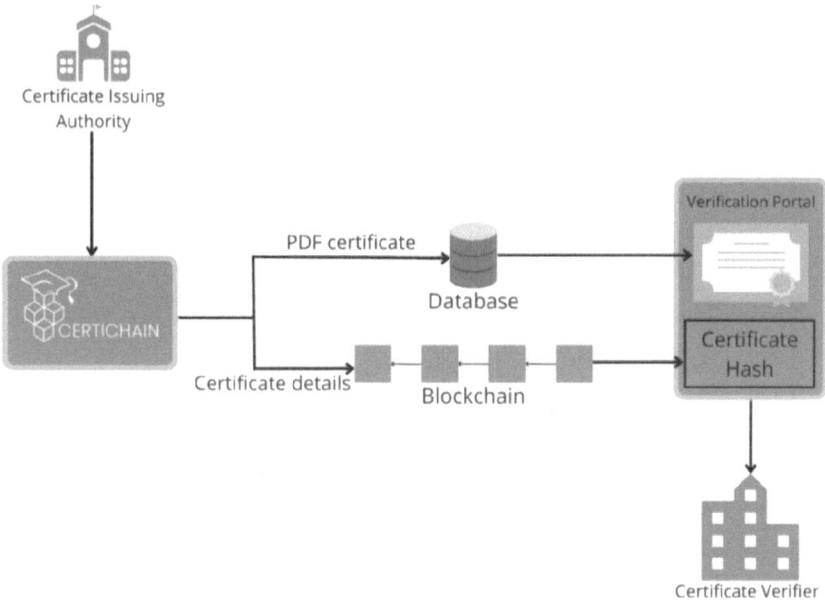

Figure 7.1 CertiChain workflow.

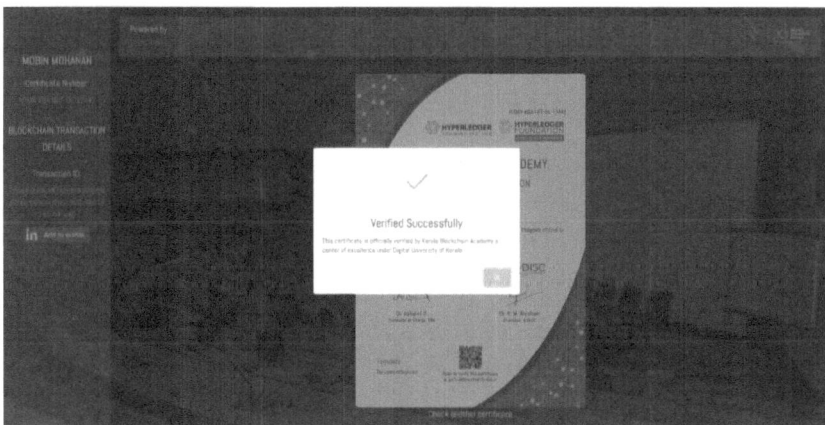

Figure 7.2 Verification portal with the verification message, issued certificate and block-chain transaction ID.

- **Batchwise Issuance:** Certificates can be issued in bulk for each batch of the course.
- **Tamper Proof:** Blockchain integration to add certificate details enables immutability and makes the certificates tamper proof, thus enabling the candidates and verifiers to be confident about the authenticity of the certificates.

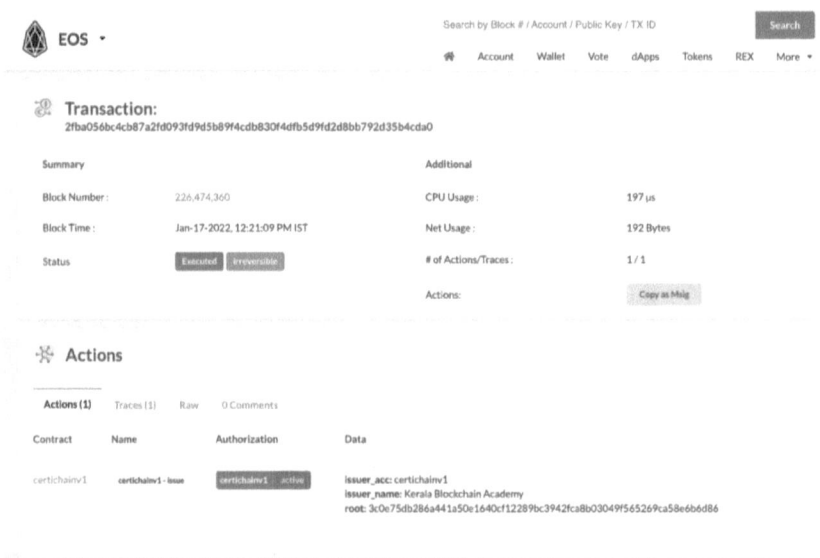

Figure 7.3 Certificate issuance transaction in blockchain.

- **No Wear and Tear:** The digital nature of the certificates prevents physical damages that may occur to paper-based certificates. The decentralised nature of blockchain also prevents the certification platform from a single point of failures.
- **Mobile Ready:** CertiChain is mobile ready and enables the verification of the certificate on the go.

The qualifications of a student who completes a course solely depend on the certificate they obtain after the successful completion of the course. The certificates are expected to be issued by institutions and organisations with proper authority. Once these certificates are issued, the verification of the same by a third party is a tedious process. So a verifier has to believe the candidate who produces the certificate and the certificate they produce. The introduction of blockchain helps in removing the requirement of trust between the candidate and the verifier. The blockchain-based certification system enables a trust-free platform for the authenticity verification of certificates. The inherent secure nature of blockchain makes sure that the certificates issued are authentic and tamper proof, thus making it difficult to fabricate certificates. Anyone with the certificate can verify that the certificate is indeed issued by a valid institution or organisation. The adoption of a paperless system will be a stepping stone in the transition toward an environment-friendly sustainable education system. This will reduce the need to print out and make copies of the certificates every time. These features combined with an easy-to-use graphical user interface help in making the certification and verification a seemingly effortless process. The QR code embedded in the certificates helps in instant verification without the long time-consuming process of certificate verification.

7.2.2 Case study 2: Sign-A-Doc – blockchain-powered multi-party document signing and verification system

Sign-A-Doc is a blockchain-powered multi-party document signing and verification system developed by Kerala Blockchain Academy, an autonomous innovation centre under Kerala University of Digital Sciences, Innovation and Technology, Thiruvananthapuram, Kerala, India. The system is developed to replace the traditional paper-based document signing and verification, thus making the document signing process efficient and environment-friendly. Managing documents often face challenges in handling and storage. The traditional way of using printed documents may result in data loss and counterfeits. Even introducing a digital system to handle documents is not enough as they are also prone to data loss and counterfeiting. The origin and integrity of documents are often hard to verify. Sign-A-Doc aims to streamline the process of signature processing, issuance and retrieval. The system helps users get multi-party signatures in the document. The blockchain underpinning ensures that the files are immutable and keep track of the state of their document in the signing process. In a few steps, users can get their documents notarised. Sign-A-Doc supports multiple page documents of numerous file formats. The system also supports multi-party signing. A QR code will be added to the documents added in Sign-A-Doc, which can be used to verify the authenticity of the document in an instant. The blockchain saves an accurate time stamp/date and the identity of the person who signed the document. Thus, Sign-A-Doc provides an immutable, paper-free, tamper-proof document signing platform, where multiple parties can collaborate.

The Sign-A-Doc system works as follows:

The Sign-A-Doc system provides an end-to-end platform for document signing and verification. It handles the processes of uploading the documents, assigning signees, notifying signees, collecting signatures, notifying the assigner, adding the transaction details and hash to the blockchain and providing the updated document to the verification portal for public verification.

The working of the Sign-A-Doc document signing and verification system involves the following steps:

Step 1: Assigner registers to the platform
    Assigners who want to get a document signed create an account and sign in to the platform. Once signed in, they can use the features

available in the system. The assigner will have the option to upload documents, select signing areas and assign signers. The assigner itself can sign the document or forward them to other signees.

Step 2: Assigner uploads document

The system supports PDFs, images and .doc files. The user can choose the file to be signed and upload them to the system. The system also supports documents with multiple pages.

Step 3: Assigner adds signees

Once the document is uploaded, the assigner can select the areas to be signed in the document and can assign signees to these areas. There is no limit to the number of signatures that can be added. The signees are added using their email address. The system also provides an option to add a custom message to these signees. The signees will get an email notification with the signing request.

Step 4: Signers sign the document

The document can be previewed and signed by the signees through the email notification received. The signers don't have to register to the system to add their signature. Their respective signatures will be highlighted, and they can sign the document by drawing their signature or by choosing an already existing one.

Step 5: Document added to the blockchain for verification

The overall working of the Sign-A-Doc system is illustrated in Figure 7.4. A notification will be sent to the assigner once all the signers sign the document. The transaction corresponding to the signing would be saved on the blockchain along with the document hash. A QR code will be added to the document for easy

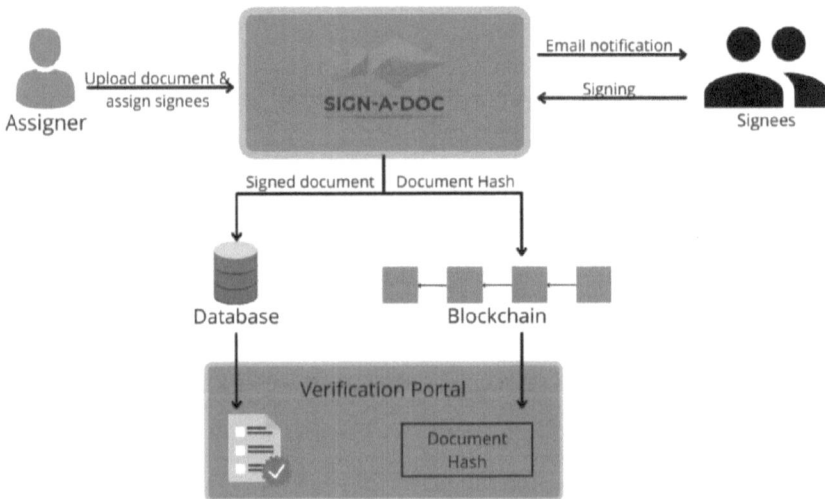

Figure 7.4 Sign-A-Doc workflow.

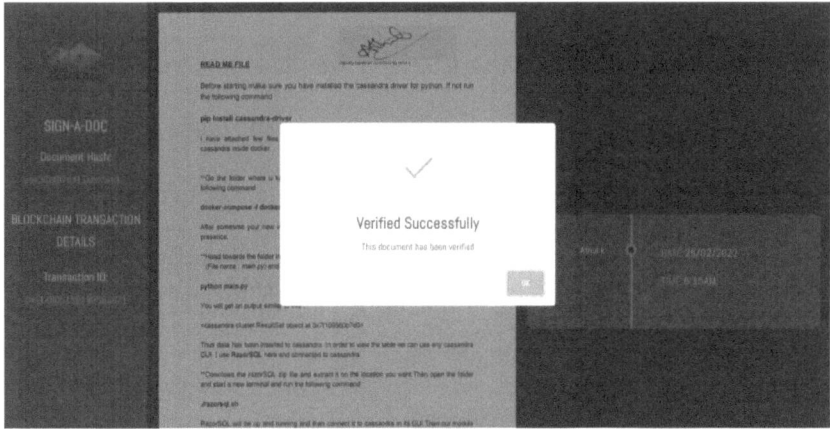

Figure 7.5 Verification portal with the verification message, signed document, document hash, blockchain transaction ID, signee name and signing details.

verification. The original signed document will be stored in a database and will be retrieved when the verification portal is used. Figure 7.5 shows the verification portal.

Once these steps are completed, anyone can use the verification portal to verify the authenticity of the document signed. The verification portal will display the signed document along with the document hash, blockchain transaction ID and signature details. The details of the transactions can be viewed in the blockchain explorer, as seen in Figure 7.6.

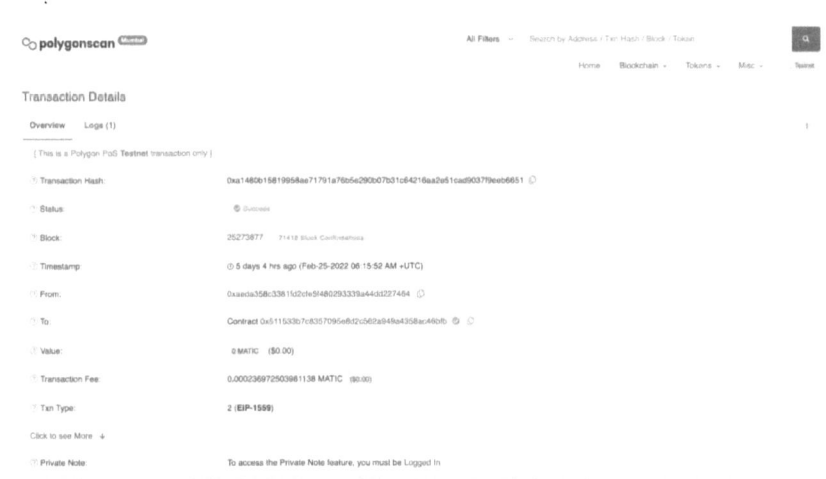

Figure 7.6 Signed document transaction in blockchain.

Following are the key features of the system:

- Streamlined document signing and verification system with the security of blockchain.
- Paperless documentation provides an environmentally friendly alternative to paper-based document management.
- Multi-party signing: Assigners can add multiple signers to the document with their email addresses.
- Multi-type and multi-page support: Assigners can upload documents with multiple pages. The system supports document types including PDFs, images and.doc files.
- Tamper-proof and Verifiable: Sign-A-Doc is backed by blockchain technology. Anyone with access to the verification portal can verify the authenticity of the document.
- Instant verification of the document through QR code.
- Authenticity verification through the document hash stored in blockchain. If the signature is made outside the application or by some other means, the authentication will fail.
- Easy-to-use graphical user interface for the assigner and signer to manage and sign their documents.
- Instant notifications to the participants in the signing process.

There are several cloud-based signing applications available in the market. They lack blockchain-based security and immutability. A blockchain-anchored signing and verification system provides an additional layer of security and traceability. Authorised parties can verify that copies of the document are legitimate by matching their version's hash with that of the original version's hash stored in the blockchain. In the case of multiple copies with different signatures, parties can use the blockchain document signing platform to look at the timestamps associated with the document's metadata. Using blockchain-based document signing systems such as Sign-A-Doc, document verification is possible anytime, anywhere.

There is no need to blindly trust third parties, neither for timestamping nor for storing signatures. It is possible to prevent anyone from manipulating the document. It is also possible to assign the signee in the system so the document remains prioritised and doesn't get shared unnecessarily.

7.3 FUTURE RESEARCH DIMENSIONS

Blockchain has been mainly used for storing academic records and transcripts. Easy verification and validation of academic credentials are made possible by blockchain solutions. Another use of blockchain is in the gamification of education, to make education more fun and engaging. Tokenisation has been used to incentivise students to achieve academic excellence. In the future, blockchain technology is expected to play a crucial role in providing a decentralised environment for collaborative student learning. Blockchain along with augmented/virtual reality, decentralised finance and tokenisation is expected to provide a sustainable learning environment for students without any boundaries or limitations. Integration of these cutting-edge technologies with blockchain will enable a sustainable education ecosystem to exist.

7.4 CONCLUSIONS

Education is one of the pillars on which modern-day civilisation is built. This chapter discusses blockchain for education and focuses on the sustainability of learning. The literature on some of the important applications of blockchain in education is covered in the review section. The use of blockchain in educational records management and incentivisation of learning is discussed. Next, two case studies on a blockchain-based certificate management system – "CertiChain" and a blockchain-powered multi-party document signing and verification system – "Sign-A-Doc" are presented. These use cases provide an example of how blockchain can be used in areas of education to improve its quality and thus help in achieving the sustainable development goal.

REFERENCES

Alammary, A., Alhazmi, S., Almasri, M., & Gillani, S. (2019). Blockchain-based applications in education: A systematic review. *Applied Sciences*, 9(12), 2400.

Antonaci, A., Klemke, R., Lataster, J., Kreijns, K., & Specht, M. (2019). Gamification of MOOCs adopting social presence and sense of community to increase user's engagement: An experimental study. *European Conference on Technology Enhanced Learning*, 172–186.

Bartoletti, M., Cimoli, T., Pompianu, L., & Serusi, S. (2018). Blockchain for social good: A quantitative analysis. *Proceedings of the 4th EAI International Conference on Smart Objects and Technologies for Social Good*, 37–42.

Bucea-Manea-țoniș, R., Martins, O. M. D., Bucea-Manea-țoniș, R., Gheorghiță, C., Kuleto, V., Ilić, M. P., & Simion, V. E. (2021). Blockchain technology enhances sustainable higher education. *Sustainability (Switzerland)*, 13(22). https://doi.org/10.3390/su132212347.

Caldarelli, G., & Ellul, J. (2021). Trusted academic transcripts on the blockchain: A systematic literature review. *Applied Sciences*, 11(4), 1842.

Chen, Y. (2018). Blockchain tokens and the potential democratization of entrepreneurship and innovation. *Business Horizons*, 61(4), 567–575.

Devine, P. (2015). Blockchain learning: Can crypto-currency methods be appropriated to enhance online learning? In *ALT Online Winter Conference 2015*.

Düdder, B., Fomin, V., Gürpinar, T., Henke, M., Iqbal, M., Janavičienė, V., et al. (2021). Interdisciplinary blockchain education: Utilizing blockchain technology from various perspectives. *Frontiers in Blockchain*, 3, 58.

Garg, A., Kumar, P., Madhukar, M., Loyola-González, O., & Kumar, M. (2021). Blockchain-based online education content ranking. *Education and Information Technologies*, 1–23.

Guustaaf, E., Rahardja, U., Aini, Q., Maharani, H. W., & Santoso, N. A. (2021). Blockchain-based education project. *Aptisi Transactions on Management (ATM)*, 5(1), 46–61.

Harnessing Blockchain for Sustainable Development: Prospects and Challenges | UNCTAD. (2022). Retrieved 6 March 2022, from https://unctad.org/webflyer/harnessing-blockchain-sustainable-development-prospects-and-challenges.

Jirgensons, M., & Kapenieks, J. (2018). Blockchain and the future of digital learning credential assessment and management. *Journal of Teacher Education for Sustainability*, 20(1), 145–156.

Kamišalić, A., Turkanović, M., Mrdović, S., & Heričko, M. (2019). A preliminary review of blockchain-based solutions in higher education. *International Workshop on Learning Technology for Education in Cloud*, 114–124.

Koh, S. (2019). *Blockchain, OpenCerts and Partnership*. Government Digital Services, Singapore. Retrieved March 3, 2022, from https://blog.gds-gov.tech/partner-us-on-opencerts-c23d38219f7e.

Kolvenbach, S., Ruland, R., Gräther, W., & Prinz, W. (2018). Blockchain 4 education. In *Proceedings of 16th European Conference on Computer-Supported Cooperative Work-Panels, Posters and Demos*. European Society for Socially Embedded Technologies (EUSSET).

Lee, D., & Park, N. (2021). A blockchain-based Untact education system for the post-COVID-19 era. *Ilkogretim Online*, 20(3).

Linux Foundation's Hyperledger Project Announces 30 Founding Members and Code Proposals to Advance Blockchain Technology – Hyperledger Foundation. (2022). Retrieved 6 March 2022, from https://www.hyperledger.org/announcements/2016/02/09/linux-foundations-hyperledger-project-announces-30-founding-members-and-code-proposals-to-advance-blockchain-technology.

Lutfiani, N., Aini, Q., Rahardja, U., Wijayanti, L., Nabila, E. A., & Ali, M. I. (2021). Transformation of blockchain and opportunities for education 4.0. *International Journal of Education and Learning*, 3(3), 222–231.

Nakamoto, S. (2008). Bitcoin: A peer-to-peer electronic cash system. *Decentralized Business Review*, 21260.

Ocheja, P., Flanagan, B., Ogata, H., & Oyelere, S. S. (2022). Visualization of education blockchain data: Trends and challenges. *Interactive Learning Environments*, 1–25.

Ocheja, P., Flanagan, B., Ueda, H., & Ogata, H. (2019). Managing lifelong learning records through blockchain. *Research and Practice in Technology Enhanced Learning*, 14(1), 1–19. https://doi.org/10.1186/s41039-019-0097-0.

Raimundo, R., & Rosário, A. (2021). Blockchain system in the higher education. *European Journal of Investigation in Health, Psychology and Education*, 11(1), 276–293.

Shah, D., Patel, D., Adesara, J., Hingu, P., & Shah, M. (2021). Exploiting the capabilities of blockchain and machine learning in education. *Augmented Human Research*, 6(1), 1–14.

Sharples, M., & Domingue, J. (2016). The blockchain and kudos: A distributed system for educational record, reputation and reward. *Lecture Notes in Computer Science (Including Subseries Lecture Notes in Artificial Intelligence and Lecture Notes in Bioinformatics)*, 9891 LNCS, 490–496. https://doi.org/10.1007/978-3-319-45153-4_48.

Steiu, M.-F. (2020). Blockchain in education: Opportunities, applications, and challenges. *First Monday*, 25(9). https://doi.org/10.5210/fm.v25i9.10654

Swan, M. (2015). *Blockchain: Blueprint for a New Economy*. O'Reilly Media, Inc.

THE 17 GOALS | Sustainable Development. Sdgs.un.org. (2022). Retrieved 6 March 2022, from https://sdgs.un.org/goals.

UntungRahardja, S. K., & EkaPurnamaHarahap, Q. (2020). Authenticity of a diploma using the blockchain approach. *International Journal of Advanced Trends in Computer Science and Engineering*, 9(1.2), 250–256.

Visa Fact Sheet. Visa.co.uk. (2018). Retrieved March 6, 2022, from https://www.visa.co.uk/dam/VCOM/download/corporate/media/visanet-technology/about-visafactsheet.pdf.

Chapter 8

Blockchain technology in banking and financial service sectors in India

Swaraj M.
Indian Institute of Management

CONTENTS

8.1 INTRODUCTION

Today, nascent digital innovations, such as online banking, phone banking, and fintech, have provided new experiences to customers. These new financial technologies (fintech) have been faster than banks in capitalising on digital technology advances, creating more user-friendly banking products that are cheaper to bring into the markets through digital means (Mohamed and Ali, 2019). The confidence surrounding fintech drove venture capital (VC) financing to a new high of US$115 billion in 2021, hitting an all-time high of $53.2 billion lines in 2018. According to the KPMG, 2021 report, the increase in transaction value in India, the United Kingdom, and the United States has fuelled fintech investment. This vast growth led to the development of new, cutting-edge financial technologies such as Bitcoin, cryptocurrencies, and blockchain.

Blockchain, a decentralised, open ledger having transaction records, is a database that every network node shares, modified by miners, supervised by all, and not owned or dominated by anyone. It operates similar to an enormous collaborative spreadsheet that everyone gets access to make it up-to-date and verify the uniqueness of digital payment transactions (Zhang et al., 2020). This technology has sparked significant investment in developed Western European countries and revamped their overall processes using this technology. Israel established the blockchain technology platform even though Russia approached blockchain technology cautiously.

India's blockchain technology initiative aims to set trustworthy online platforms based on common blockchain facilities by boosting research and innovation, development, techniques, and integration systems. It enables citizens and businesses to receive advanced, reliable, stable, and credible digital services, cementing India's leading position in blockchains. All these resulted in rapid development and enormous opportunities for India's socioeconomic sector as it generates large numbers of jobs and boosts technology-based economic growth. India has announced its blockchain-based currency, a digital currency operated by the Reserve Bank of India, by 2023, thereby accelerating legalising and regulating digital assets.

Thus, the chapter systematically describes blockchain technology and its components, types, characteristics, strengths, weaknesses, opportunities, and threats in India's banking operations and financial services.

8.2 BLOCKCHAIN TECHNOLOGY

Blockchain is still relatively new to the market as an ingredient of the fourth industrial revolution. This technology was used for the first time in 2008 when (Nakamoto, 2008) a white paper was written about Bitcoin, a digital version of money that could make electronic transactions among both people without the help of centralised intermediaries in finance. Nakamoto developed a ledger to encourage this digital money that he named "a chain of blocks"; afterwards, he dubbed it "Blockchain." This subject started gaining traction only in 2016 after introducing Bitcoin trading and embracing bitcoins as a form of virtual transaction in fast-growing markets. Today, blockchains extend their applications from sector to sector.

Blockchain is a conglomeration of distinct advancements with demonstrable financial importance. It facilitates the generation of a blockchain network that provides a unified view for all parties involved in business transactions. With the use of the blockchain network, this encrypts all exchanges and associated data using hashing, ensuring that every user possesses entry to all the corresponding databases (Zheng et al., 2017). According to Swan (2015), blockchain is a decentralised repository of constantly modified systems accessible to account holders for verification and documentation related to a database in a computer system. This technology protocol permits the protected exchange of money and securities and user authentication without the involvement of a third-party financial institution. Additionally, it secures payments involving virtual currencies and user information about equity, debt, digital assets, and copyrights. Since each decentralised entity must approve stored data individually, it cannot be easily forged or tampered with. Thus, blockchain technology establishes trust and reduces accounting costs prevalent in emerging markets and relevant activities. This technology is a protocol for exchanging value without using an intermediary over the Internet.

8.3 COMPONENTS OF BLOCKCHAIN TECHNOLOGY

The four major components of blockchain are

1. **Hash:** It assigns unique indexes using one-way mathematical functions.
2. **Digital Certificate:** It takes the form of a public encryption code.
3. **P2P Application:** It serves as a data transmission configuration for nodes to connect the decentralised hash.
4. **Crypto Contract:** The crypto contract is a set of integrated techniques which guarantees reliability, quality, and precision of stored information throughout all network participants.

8.4 TYPES OF BLOCKCHAIN TECHNOLOGY

The different kinds of blockchains are

- **Public Blockchain:** It is a completely open and peer-to-peer technology. Anyone can read, send, and be a part of the message exchange process.
- **Permissioned Blockchain:** It is a quasi-decentralised technology, wherein consensus is controlled by a few pre-selected nodes and read access is restricted to participants only.
- **Private Blockchain:** It is centralised technology that requires a 'high trust' entity with write permissions granted to a single entity and read permissions granted to all participants.

8.5 CHARACTERISTICS OF
BLOCKCHAIN TECHNOLOGY

Blockchain technology is advancing rapidly and is updating its features day by day. The unique characteristics of blockchain technology are

- **Transparency:** Enabling information transparency is the most important promise of blockchains, which creates an entirely auditable and valid ledger of transactions. The blockchain is intended to be a transparent machine where anyone can join and view the network's data.
- **Data Integrity:** Blockchain technology protects against data manipulation by being immutable once added to the shared ledger. It enables a higher level of data traceability and auditability, authorising for outlining any data entered incorrectly earlier to the agreement.
- **Immutability:** Immutability refers to the inability of something to be changed or altered. It is one of the essential blockchain features because it enables the technology to remain a permanent, unalterable network.
- **Privacy:** Blockchain technology produces a data structure with inherent privacy qualities. Cryptography ensures the privacy of the records on a blockchain. Each private key is associated with each transaction and acts as a digital signature for the respective participant. Notification promptly prevents any damage.
- **Reliability:** It is trustworthy because each access point keeps a replica of the blockchain database. Thus, even if each entity becomes unavailable, the record remains accessible to the remaining peer nodes.
- **Versatility:** The versatility of blockchain technology enables businesses to collaborate safely with business partners in a protected environment. It made its way into numerous areas, demonstrating its adaptability in various scenarios (Figure 8.1).

Blockchain technology can be used in many different fields, including financial services, governance, healthcare, education, value chain, network security, banking sector, communications, regulatory, energy market, and so on. Many people worldwide and in different countries are trying to use distributed ledger apps, especially in banking and financial sectors.

8.6 USING BLOCKCHAINS IN
BANKING AND FINANCE

Blockchain technology has grown in importance as one of the most powerful and promising banking and financial services technologies (Kim et al., 2020). Large corporations are primarily involved in traditional banking, large-scale borrowing, insurance, personal loans, mortgage finance,

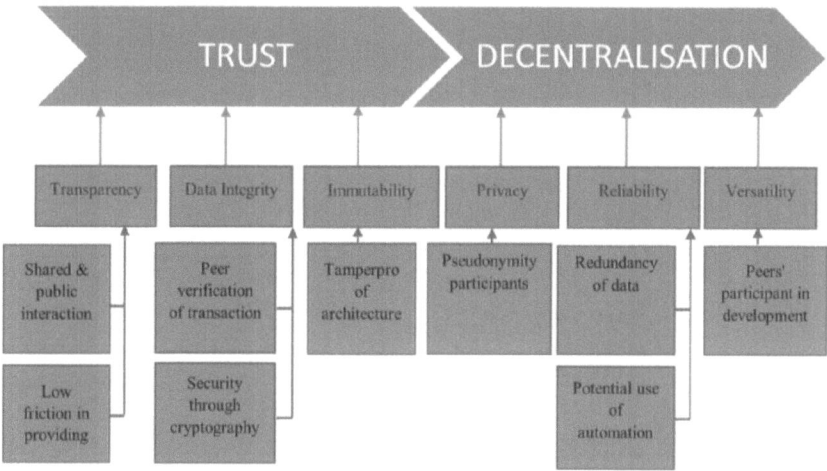

Figure 8.1 The major characteristics of blockchain technology.

reputation management, and wealth management. Blockchain technology is being applied to a variety of banking and finance services provided by businesses, including payment processing, financial instrument issuing, financial consolidation and peace deal, cross-border payouts, and cashless transactions (Schuetz and Venkatesh, 2020; Zachariadis et al., 2019) (Chang et al., 2020). Undoubtedly, such blockchain interventions will lead to radical changes in the banking and financial sector and a new kind of leap forward.

Financial technologies have catalysed financial market advancement, with blockchains playing a critical role in this modernisation (Lee and Shin, 2018). Blockchain technology, categorised into five principles, including logical programming, peer-to-peer communication, avoidance of documentation, transmitted database, and username visibility, has enormous potential to free and regenerate the financial service sector (Tapscott and Tapscott, 2018). It provides more choices to small- and medium-sized capital market participants, making the service more accessible and personalised. These financial systems include derivative instruments, asset securitisation, peer-to-peer lending, and fundraising (Figure 8.2).

While existing financial services range from transactions to funds, blockchain technology upended this model by enabling more pioneering, secure, and efficient exchanges at a cheaper cost (Lee and Shin, 2018). Similarly, these tools enable the banking sector to obtain users' spending ratios, investing details, and even their personal expenses via a multifunctional integration platform. This assists in overcoming insufficient data dealing with financial customers to obtain credit. Moreover, in a decentralised system, a blockchain network acts as an intermediary for exchanging assets and information.

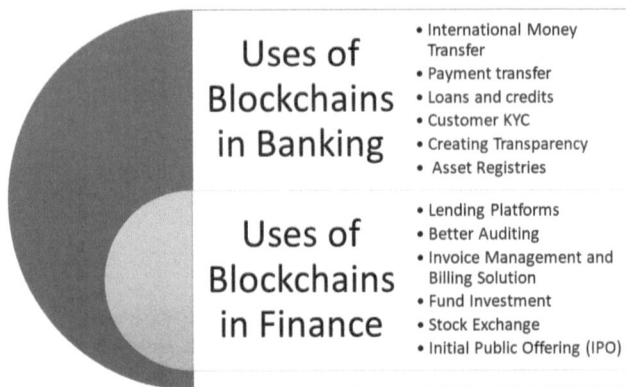

Figure 8.2 Uses of blockchains in banking and finance.

Blockchain technology also implements point-to-point payment that obviates the need for a third party as an intermediary. It significantly improves operational effectiveness and cuts down the payment system. As a result, banks can meet the growing demand for quick and simple bank transfer settlement services. Furthermore, blockchain technology helps to handle a complex network of accounting systems, such as record-keeping, payment reconciling, adjustment settlement, and account preparation. Likewise, banking services ensure fraud prevention, data security, money laundering prevention, automation and standardisation, real-time information sharing, and the development of predictive models using blockchain technology. Blockchain technology is regarded as the next upheaval that will alter the banking and financial industry's structure and size and how corporate transactions are done (Cermeno, 2016).

Table 8.1 exhibits the comparison of traditional banking, online financial services, and cryptocurrency banking.

Table 8.1 Comparing traditional banking, online financial services, and cryptocurrency banking

	Traditional banking businesses	Online financial services (FinTech 1.0)	Cryptocurrency banking (FinTech 2.0)
Customer experience	Identical contexts	Diverse contexts	Diverse contexts
	Unified service	Customised service	Customised service
	Poor customer experience	Good customer experience	Good customer experience
Productiveness	Numerous connections to middlemen	Numerous connections to middlemen	Transmission from node to node, no middleman

(Continued)

Table 8.1 (Continued) Comparing traditional banking, online financial services, and cryptocurrency banking

	Traditional banking businesses	Online financial services (FinTech 1.0)	Cryptocurrency banking (FinTech 2.0)
	Intricate settlement procedure	Intricate settlement procedure	Shared ledger, payment= settlement
	Low productiveness	Low productiveness	High productiveness
Cost	Multiple conventional audits	Small conventional audits	Digitalised
	Numerous connections to middlemen	Numerous connections to middlemen	No middleman
	Cost increases	Cost increases	Cost decreases
Security	Centralised storage space is susceptible to encroachment.	Centralised storage space is susceptible to encroachment.	Centralised storage cannot be encroached upon.
	Easy to access users' individual data	Easy to access users' individual data	Encryption codes to protect user's individual data
	Insufficient security	Insufficient security	Sufficient security

8.7 SWOT OF BLOCKCHAIN TECHNOLOGY IN BANKING AND FINANCIAL SERVICES

8.7.1 Strengths

Blockchain technology facilitates stakeholders to store data decentralised, diligent, irreversible, and consensus-based. It enables efficient storage of ledgers in a distributed environment. It also permits for proper verification and confirmation relating to the role of the stakeholder. Moreover, depending on the application domain's requirements, either a permissioned or permissionless blockchain environment may be configured. It helps seamless integration of blockchain technology into various application domains as a service.

The following are the significant strengths of blockchain technology.

- **Security:** By granting users ownership of their records and stopping subsequent entities from misrepresenting or acquiring them, blockchains can boost privacy and security.
- **Transparency:** It is increased as a result of payments being distributed throughout the system.

- **Enhanced Privacy:** In comparison to monopolising all information in a centralised data system and preventing database loss from threats, blockchains improve the security of financial institutions.
- **Performance:** Banking institutions can rapidly reduce their overhead costs using blockchain solutions.
- **Immutability:** The blockchain's transaction history cannot be altered because the blockchain ledger is immutable and unalterable.
- **Quick Transactions:** Transactions will be able to reach participants more quickly even with small steps thanks to the blockchain.
- **Reliance:** Blockchain technology is based on collective action, which ensures that every participant adheres to previously agreed-upon rules.

8.7.2 Weaknesses

Each technology has drawbacks, but early identification of these flaws enables preparation to address issues well in advance of them affecting the emerging blockchain network. The given considerations must be made prior to implementing blockchain-based systems.

- **Extensibility**

  In relation to 1,000 existing financial operations per second, blockchains can be processed at a rate of approximately five per second, based on the system's applicability to a specific area, structural considerations, general agreement view, and the size of the network operated, among other factors. Additionally, the basic structural solutions like hyper ledger texture appear to be acceptable for demands from a number of domains, whereas researchers continue to focus on scalability and strive to achieve better results. This discrepancy creates a significant barrier to the global adoption of blockchain for banking.

- **Security**

  Contrary to popular belief, all weaknesses are not universally applicable to all blockchains or the corresponding emerging applications. It is necessary to develop a security risk model similar to threat perception models including scamming, malicious attacks, denial of service (DoS), information leakage, abrogation, and acceleration of benefit. Attempts are currently being made to create an online repository of known blockchain data breaches. Hackers can acquire confidential data by interpreting data or adjusting financial statements.

- **Interworking**

  Interoperability between diverse blockchain networks is still a work in progress, and much research needs to be done to solve this gap. The situations that require interworking are as follows:

  1. When an organisation migrates to a modern blockchain system, it must maintain compatibility with cloud-based platforms.
  2. When users choose disparate blockchain systems, they must communicate with one another.

3. When a platform's retention limit is exceeded, a proposed system must be established to communicate with the previous platforms.

There is a lack of interconnectivity among various blockchain systems, which is pivotal for financial companies to make payments.

- **Data Localisation**

  Data localisation entails collecting, storing, and processing data about the citizens or residents within the country prior to being transferred or shared internationally. Countries have enacted new data laws to restrict data flow and localise data. In the European Union, the General Data Protection Regulation law was enacted. Under certain conditions, the Personal Data Protection Bill suggested by India would enforce the gathering, backup, and handling of individual information as well as their exchange to other countries.

- **Record Disposal**

  It is one of the rights proposed by the Personal Data Protection Bill. Due to the immutability of blockchain records, appropriate measures must be taken to enforce this requirement during the implementation of blockchain technology.

- **Reversibility**

  Financial companies cannot reverse an incorrect blockchain payment as transfers are permanent and last.

- **Regulatory Threats**

  Blockchain technologies (for example, in transactions, loans, and investing money) are still in their infancy and experimental stages in the finance sector, confusing and testing regulation.

8.7.3 Opportunities

Any new innovative technology introduces a variety of opportunities that significantly impact the implementation stage.

- Blockchain technology improves the secure storage of categorised and uncategorised records. As a result, this potential is appropriate for all those who require it.
- By virtue of its end-to-end potential, blockchain enables clear transaction processing across all domains.
- The concept of blockchain technology facilitates the evolution of value propositions for a wide range of options (B2B, G2G, G2C, etc.).
- **Advantages in the Marketplace:** Blockchain technology gives banks long-term advantages in terms of financial transactions because it cuts costs, makes information more accessible, and helps them better manage risks.
- **New Service Formation:** In the future, blockchain technology will assist banks in developing new services.

8.7.4 Challenges

Regulatory issues are one of the most significant critical challenges that blockchain technology faces, and they are one of the severe challenges in general. However, the varying uses of technology necessitate procedures such as digital currencies, cryptocurrencies, and the adoption of blockchain-based systems. Therefore, proper regulatory methods must be implemented depending on the economic operations such as transactions, borrowings, and investments provided by the blockchain. This regulatory issue leads to the use of blockchains to minimise taxes and other nefarious offences. Thus, regulatory approaches must strike a delicate balance between the innovative spirit of the technology and the chance that technology contributes accidentally to measurable threats in the banking system (Yeoh, 2017). The major challenges of blockchain technology are given in Figure 8.3.

- **Adoption of New Technologies**
  It is imperative to conduct a systematic review of return on capital, compliance, safety and confidentiality, and efficiency when determining the applicability of blockchain in a network domain with the fast expansion of blockchain applications getting constantly established.
- **Adherence to Rules**
  Although blockchain technology is proposed to be used in a specific usage arena, this is critical to research adherence to law and any consequences for the usage arena. Therefore, additional regulatory policies have to be developed in response to the requirements.

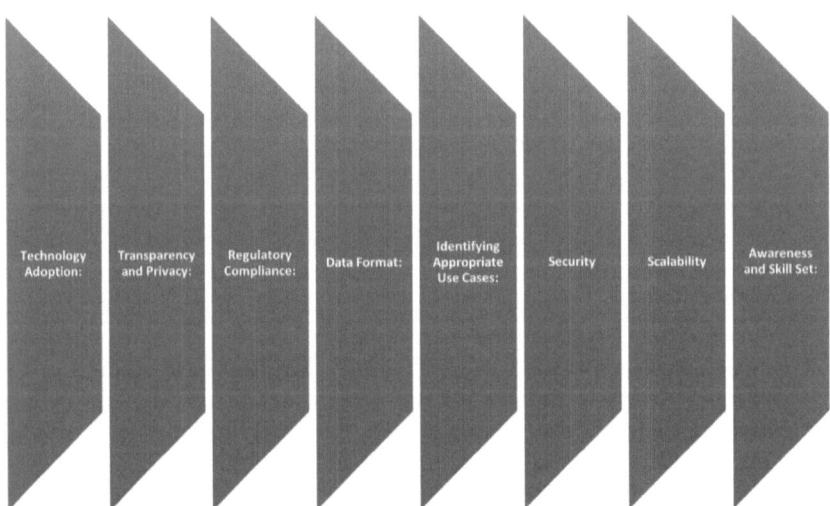

Figure 8.3 Challenges of blockchain technology.

- **Identifying Appropriate Applications**

  Different applications require varying degrees of security, privacy, and data storage, depending on the number of participating entities. Thus, the suitability of blockchain in a given application context must be carefully evaluated.

- **Data Format**

  Success in implementing blockchain's functionalities depends on whether the payment format has been identified in various co-environments or on adhering to its associated characteristics, such as its insistence on other data, among other factors.

- **Intelligence and Skillset**

  A skilled workforce that understands the potential and relevance to the specific application arena of blockchain technology is needed to implement the system properly.

- **Openness and Confidentiality**

  Though solid encryption and rationale are crucial considerations of blockchains, they ensure that people remain virtually untraceable and process payments across addresses created by themselves. It protects people from disclosing their true selves, making the technique incredibly secure and dependable for users. This, in turn, creates some limitations to the technology. Significant numbers of user identities are permissible to avert any data leaking (Zheng et al., 2018). Unfortunately, it has been demonstrated earlier that blockchain technology does not permanently preserve payment privacy as public transactions are visible to the public.

- **Security**

  Blockchain has a plethora of uses and applications in the banking sector. In spite of these numerous benefits, the blockchain application has been impeded by its close association and recognition with cryptocurrency in the faces of public entities and legislators. This technology is notorious for its association with numerous Bitcoin frauds and scandals in the blockchain demand. These frauds and deceptions have impacted blockchain technology's reputation and trustworthiness in the eyes of regulators and societies worldwide. Although blockchain systems are inherently resistant to tampering and fraud, specific systems might still be subject to assaults and fraud attempts when multiple network members collude (Yeoh, 2017). Eyal and Sirer (2014) revealed that a trivial proportion of computing performance might be leveraged to trick a blockchain platform successfully. These safety issues could jeopardise the broad acceptance of blockchain services.

- **Adaptability**

  The payments are increasing at an alarming rate, squeezing the life of existing blockchain networks. Due to the blockchain's configuration, each activity must be secured for verification and current track. This issue is supported again by the payment processing constraints imposed by the cryptocurrency Bitcoin. Due to the block size and duration constraints, these techniques will be ineffective at extracting a large number of transactions in a short period. There is an urgent need to enhance blockchain's transactional capability. This scalability issue must be addressed for blockchain's potential benefits to be realised entirely in the banking and financial sectors.

8.8 OTHER MAJOR BLOCKCHAIN CHALLENGES IN BANKING AND FINANCIAL SERVICES

Prudential concerns are another major challenge faced by blockchain technology. If various organisations depend on the same source, execution, or collaboration of blockchain, it causes prudential risks. Additionally, the business strategic model of blockchain technology improves user value by facilitating previously inaccessible services or products. It is possible only by devoting a significant amount of time or money. This is a big challenge for blockchains. Similarly, controversies and user awareness, public regulations, constraints on access, the energy consumption of technology, uncertainty about blockchain networks, conceived initial adoption risks, lack of interconnection and segmentation, interruption of current market techniques, inadequate evidence regarding business growth and wider economic effects, inconsistency in technology governance, and process redesign and legacy system conglomeration are included in the major challenges of banking and financial services.

8.9 CONCLUSION

This study emphasises the critical importance of thoroughly comprehending the diverse perceptions of expenditure, advantages, threats, and potentials associated with developing blockchains for the banking and finance industries. Blockchain follows steam engines, electricity, and the Internet as a digital transformation. Steam engines and electricity considerably enhanced the work efficiency (Hämäläinen and Inkinen, 2019). The Internet revolutionised data transmission. Likewise, blockchains alter how money is spent to establish trust.

The prior financial industry goods and services were deemed prohibitively expensive and inefficient. With time, financial services and blockchain technology have evolved to meet the different demands of interested parties. As an emerging paradigm with enormous potential for various stakeholders in the banking and financial services, blockchain technology can completely transform this sector.

Blockchain technology has addressed two critical issues in the digital economy: The chain enables the flow of all digitalised assets. Another thing is that genuine zero-cost trust is established under an invisible society, allowing for unique possibilities. Even though the decentralised feature of blockchain benefitted financial and banking services, its application and development have been hampered by numerous obstacles. As a result, blockchain technology poses a dual threat to conventional global financial systems. Blockchain technology places a more significant burden on policymakers' data handling and threat management abilities. To serve the entire society better, we need to investigate the application tiers for the blockchain financial and banking sectors from the perspectives of new tech, adaptability, legislation, and governance.

REFERENCES

Cermeno, J. S. (2016). Blockchain in financial services: Regulatory landscape and future challenges for its commercial application. *BBVA Research*, 1–33.

Chang, V., Baudier, P., Zhang, H., Xu, Q., Zhang, J., & Arami, M. (2020). How Blockchain can impact financial services – The overview, challenges and recommendations from expert interviewees. *Technological Forecasting and Social Change, 158*, 120166.

Guo, Y., & Liang, C. (2016). Blockchain application and outlook in the banking industry. *Financial Innovation, 2*, 24.

Hämäläinen, E., & Inkinen, T. (2019). Industrial applications of big data in disruptive innovations supporting environmental reporting. *Journal of Industrial Information Integration, 16*, 100105.

Kim, S., Park, H., & Lee, J. (2020). Word2vec-based latent semantic analysis (W2V-LSA) for topic modeling: A study on blockchain technology trend analysis. *Expert Systems with Applications, 152*, 113401.

KPMG. (2021). *it home.kpmg/governance.* https://assets.kpmg/content/dam/kpmg/xx/pdf/2021/08/pulse-of-fintech-h1.pdf.

Lee, I., & Shin, Y. J. (2018). Fintech: Ecosystem, business models, investment decisions, and challenges. *Business Horizons, 61*, 35–46.

Mohamed, H., & Ali, H. (2019). *Blockchain, Fintech, and Islamic Finance: Building the Future in the New Islamic Digital Economy.*

Nakamoto, S. (2008). Bitcoin: A Peer-to-peer electronic cash system. *Decentralized Business Review*, 1–9.

Schuetz, S., & Venkatesh, V. (2020). Blockchain, adoption, and financial inclusion in India: Research opportunities. *International Journal of Information Management, 52*, 101936.

Swan, M. (2015). *Blockchain Blueprint for a New Economy* (Vol. 1).

Tapscott, D. T. A., & Tapscott, A. (n.d.). *Realizing the Potential of Blockchain A Multistakeholder Approach to the Stewardship of Blockchain and Cryptocurrencies.* World Economic Forum.

Yeoh, P. (2017). Regulatory issues in blockchain technology. *Journal of Financial Regulation and Compliance*, 25, 196–208.

Zachariadis, M., Hileman, G., & Scott, S. V. (2019). Governance and control in distributed ledgers: Understanding the challenges facing blockchain technology in financial services. *Information and Organization*, 29, 105.

Zhang, L., Xie, Y., Zheng, Y., Xue, W., Zheng, X., & Xu, X. (2020). The challenges and countermeasures of blockchain in finance and economics. *Systems Research and Behavioral Science*, 37, 691–698.

Zheng, Z., Xie, S., Dai, H., Chen, X., & Wang, H. (2017). An overview of blockchain technology: architecture, consensus, and future trends. *2017 IEEE International Congress on Big Data (BigData Congress)*, 557–564.

Chapter 9

Legal dimensions of blockchain technology

Renuka V.
Cochin University of Science and Technology

Shanmugham D. Jayan
Vijayaraghavan and Devi Associates

CONTENTS

9.1 INTRODUCTION

Technology is, basically, identifying the possibilities of science and utilizing the same for simplifying human activities. If the benefit of the same exceeds the cost involved, it gets accepted in the society. At the same time, the mere social acceptance of some technology does not corner an altruistic shade. In a given system, there are multiplicities of conflicting interests. Technology is also similar in line. The acceptance factor need not mean that it is only having a positive utility, but it may have negative aspects as well. The determination of the same and its permitted extent depends on the decision of law. Law, unlike science, is not an analysis of causative factors but a determination of what should happen (Markey, 1983). For instance,

> Your interest in crossing the road straight from your place of work to the vehicle parked on the opposite side is conditioned by law and you are forced to walk about fifty meters to cross the road through the provided zebra line.

Here, the demand to cross the road through the zebra line is in furtherance to larger interest which is in variance with your interest. This is a simple example of conflict of individual interest vis-a-vis larger interests.

DOI: 10.1201/9781003282914-9

The target of law is basically balancing these two (Gardner, 1962). In the realm of technology also, this aspect is predominant. Law is not always wholeheartedly welcoming technology. Law needs to take into consideration the impact of any given technology on multiple factors. Blockchain is a technology which is of nascent origin and per se is lacking a previous counterpart; in addition, it is affecting multiple aspects of the existing system (Salmon and Myers, 2019). In such a context, blockchain is of concern to law. Since it is affecting the multiple aspects of the system, it indicates that the laws in such aspects also need to be equipped to tackle the challenges.

The core attributes of blockchain are transparency, control, and security (Zhang et. al., 2019). Blockchain, by the nature of technology itself, imbibes the feature of traceability and hence confirms transparency (Idrees et. al., 2021). The blockchain structure is capable of controlling access to the same in the context of permitted systems or networks as well as identified users. As it is impossible to alter or tamper data once entered, blockchain assures security. In other ways, this distinctiveness raises eyebrows of the legal fraternity. It affects multiple branches of law due to its implication in various existing utilities. Hence, this paper chapter analyses the intricacies of blockchain technology in the legal realm by examining its implications and challenges to the established principles of law in different sectors covering privacy, data protection, taxation, jurisdiction, property laws, and contractual relations.

This chapter is basically for narrating the impact of blockchain technology from a legal perspective. Technology plainly speaking is science in action. Science is basically realization of casualties by the phenomenon of observations and rationalizing of the same. This identification of casualties and utilization of the same, for the purpose of achieving something of utility to human interactions, is achieved by technology. The element of involvement in societal interactions elevates it to the level of a concern of law. It is an accepted fact that law lacks technology and hence logically there is legal vacuum in the evolution of a technology. This chapter discusses the specific areas wherein blockchain technology creates concerns for law. It further discusses the responses and possibilities to overcome these concerns. This chapter is structured for this purpose. Section 9.2 is labeled Blockchain Technology: A Socio-Political Overview; the same shall give an overview of the nature and scope of blockchain technology. Even though a technological elaboration is required to analyze the nature and scope of blockchain technology and public distributed ledger (Yaga et al., 2019; Reyes, 2017), this part conveniently avoids the same and gives an account of socio-political dimension. Instead, this part covers the developments and concerns of that period in concise. Section 9.3 shall cover the legal dimensions of the technology elaborating the conceptual and practical challenges of the legal system to accommodate the same into its existing realm covering a few areas in specific and is thus given the heading Legal Repercussions of Blockchain Technology. Section 9.4 is the Conclusion.

9.2 BLOCKCHAIN TECHNOLOGY: A SOCIO-POLITICAL OVERVIEW

In logical terms, emergence of the art of cryptography is clubbed with the evolution of means of communication (Paar and Pelzl, 2009; Bauer, 2002; Kahn, 1996). The legal realms of the same nevertheless are of recent origin. The second half of the 20th century witnessed the geometrical development of information and communication technology (ICT hereafter) and the conversion of the art of cryptography into a pure mathematical phenomenon (National Academies of Sciences, Engineering, and Medicine 2018; Zapechnikov, Tolstoy, and Nagibin, 2015). The self-achieved capacity of the subjects for secretly communicating also became a concern for the state and hence triggered the development of law of cryptography. Without any doubt, cryptography is a means for improving the concept of privacy, and hence the same is also a barrier for the state in its monitoring of communication of the subject (Shearer and Gutmann, 1996). The fact that those who are acting against the state opt for cipher messages is the factor that is compelling the state to control cryptography (National Research Council, 1996). The prevalence of a comparatively high degree of individual liberty coupled with the scope for technologies made the United States of America a fertile jurisdiction for laws in cryptography (Lerch and Gray, 1997). Cryptography was brought into the net of the Arms Control Act/Defense Trade Regulations (formerly known as the International Traffic in Arms regulation – ITAR) (Radlo, 1997). Being equated with arms, there were restrictions on export or import. This could be cited as the first instance of legal control of cryptography.

Another significant development was in the form of clipper chips, wherein the US authorities demanded major telecom companies to share the cryptographic algorithms used in telephonic instruments (Froomkin, 1994). This was a compromise between the state and the subjects. The state limited the subject's options of cryptographic use through the regulation of the complexity of algorithms. As usual, there was a conflict between the interests of the state and subjects. As a part of the compromise, freedom for using cryptographic tools was mooted. However, technology developers were mandated to share a copy of the said algorithm to an independent statutory body. This body needs to share the copy provided with the government for monitoring. This body is identified as a trusted third party for this purpose (Froomkin, 1996).

The development of public key cryptography is another milestone in the technical as well as the legal conundrum (Froomkin, 1994). The difficulty in breaking open a cipher, produced using public key cryptography, was a big achievement in the technical arena. At the same time, it sets another serious challenge to the law makers. A very practical utilization of public key cryptography was in the realm of implementation of digital signature (Kaur and Kaur, 2012). In the said process, carrying the idea of a trusted third party of the crypto chip scenario emerged certifying authority who were licensed by the government to issue key pairs and in turn digital signatures to the subscribers. Here also, even though it can be seen that public key cryptography is foolproof, the generation of trust is not merely because of the technology but because of the presence of the trusted third party. The legal recognition of the third party and the realistic level of verifications done by the certifying authority resulted in public key infrastructure and the element of trust on the same.

Blockchain technology is also a byproduct of growth of cryptography. However, it was having certain definite advantages above and over the pure use of public key cryptography (Dong, Luo, and Liang, 2018). In the realm of blockchain technology, the requirement of a trusted third party or certifying authority is not required. Any one contributing to the block is never in a position to absolve from the ownership or involvement in the chain. The trust thus is inherently part of the blockchain technology which to a great extent was induced in earlier forms of implementation of cryptography. The nature of blockchain itself is ensuring nonrepudiation and security of the information obtained therein. The reliance of cryptography is definitely there because it is the said technology which is getting successfully utilized. All the legal concerns that were part and parcel of cryptography logically will continue in the realm of blockchain technology.

Philosophically, the completeness of blockchain technology in creating trust inherently opens new challenges to the political state as well as law. There is no need for any third party or even a legal system for the purpose of generation of trust among and between the participating entities when they are part of a closed blockchain (De Filippi, Mannan, and Reijers, 2020). Blockchain technology itself is ensuring the required trust and even more of it for the participating entities to move ahead. This is the larger dimension of blockchain technology that is capable of making the same elusive to the tentacles of law.

Specific utilization of blockchain technology in various contexts is capable of bringing in concerns hitherto unfaced by the legal system (Idrees et al., 2021). Such specificity is something which is above and over the earlier mentioned philosophical version of blockchain technology. Since there are multiple spheres in which blockchain technology can be used, there is no dearth for the resultant legal concerns that are also traversing branches of law that were hitherto dealing with the areas in which blockchain is getting implemented. There is a list of such subjects of law area as mentioned earlier, multiple in numbers.

9.3 LEGAL REPERCUSSIONS OF BLOCKCHAIN TECHNOLOGY

The societal impact of technological advancements squarely affects the legal system developed in tune with the time and place. The distinctiveness of the blockchain technology has uplifted the communication means to another level. This is, once, beyond the foreseeability of the legal system. The legal system has been undergoing changes in tune with the demands of society from time to time. The ICT-induced developments have opened new vistas of interactions by minimizing human interventions. This has a huge impact upon the system established for and by the human community within the inherent limitations. This urges for revamping the traditional- or human-induced social system so as to cope up with the technology-oriented one.

This part of the paper shall discuss the core areas of law on human interactions and transactions in the public realm related to blockchain technology. As the general impact of ICT on the existing legal system has already been in the limelight of discussions, debates, and the socio-legal considerations of the nation states, this part is annotating the same. Rather, an analysis of the legal challenges posed by the blockchain technology is portrayed.

9.3.1 Jurisdiction, Privacy, and Data Protection

The concept of jurisdictional issues is always a part and parcel of cyberspace. The same is always a misnomer with least regard to the spatial dimension of geographical boundaries of nation states to fix the legal jurisdiction. The legal community accepted and discussed this aspect in a full-fledged manner already (Menthe, 1997; Trachtman, 1998; Lemley, 2003; Hildebrandt, 2013). At the same time, the blockchain technology poses certain newer jurisdictional challenges, inherently connected with the technique used in it (Werbach, 2018). The build-up of the blockchain database perpetuates these jurisdiction concerns. Rather, more specifically, the jurisdictional concerns in the realm of blockchain technology are more or less connected with the concerns in the areas of privacy as well as data protection.

As mentioned earlier, the technique of cryptography itself is a facet trying to further privacy (Kulhari, 2018). The concern of law is that technical realm is continuing in the privacy aspects connected with blockchain. The genesis block is having the linkage to its creator; however, the conversion of the genesis block and the resultant blockchain is bringing in more participants into the same and the technical aspiration is only on the secrecy and nonrepudiation factor (Zhang et al., 2019). This technical inclination sounds good in technology but not for law. The legal realm has concerns upon the way in which the attributes of privacy are ensured in the working of such a technology. The contributors or participants as the case may be in the blockchain are actually marking their presense in the same. Such markings are also coming within the category of personal information. The concept of personal information and its various degrees of sensitivity brings in different grades of legal regulation (Schwerin, 2018). The technology used in blockchain is *per se* not concentrating on this realm, and the scope of bringing in laws on privacy is a difficult task as well. Lately, the concept of consent has been used for sanctifying many of the legal prohibitions which might otherwise have been present. However, international law through its consensual jurisprudence has elevated so many individual rights to the realm of inalienable privileges (Moore, 2018). There are aspects of such rights which are taken from the so-called basket of privacy.

In addition to this, the legal idea of consent is inherently having the adjective "informed" which requires a subjective satisfaction to the involved person upon the nature of consent provided (Faden and Beauchamp, 1986; Sprung and Winick, 1989). In an activity revolving around a high-technology application, the 'consent' that can be inferred from the activity of the contributor to the blockchain is prone to many challenges. The degree of information perceived with respect to the contributor or participant in the activity may definitely be having information that is part of the law on privacy.

The element of protection of privacy in the context of information in electronic form is taking the concept of data protection laws that are very much part and parcel of many of the major nation states (Data Protection and Privacy Legislation Worldwide) (Kulhari, 2018, 23–37). The aim of such laws is to ensure the concept of privacy of its subjects by bringing in conditions upon collection, processing, transfer, and use of data related to the subjects. The scope of the protected data is very much of a wide amplitude, and even data which are capable of identifying a person itself need to be protected.

On the other hand, the sensitivity of the data, be it personal, medical, financial, or whatever, gets a higher degree of protection as well as consequences for violations. As mentioned earlier, the concept of consent has a major role in the context of these laws. However, many times, the consent is not specific and by implication without the knowledge of the consequence. In such a situation, it is doubtful whether the said consent can be considered a legally accepted consent. Taking any standard data protection laws, the same is conceived based on the roles of different involved parties identified from the present technological scenario. Thus, based on the same, there are two specific roles of data controllers and data processors. These two roles are very crucial.

In a very conventional way of data handling, it is easy to identify and fix these roles. However, in the realm of blockchain technology, it is difficult to define and identify the entity engaged with the same. There are no participants in a blockchain technology who can be identified as falling into these roles specifically. This issue can be overcome by bringing in some variance of data protection law suitable in the realm of blockchain technology.

Nevertheless, such a demand itself is a challenge for law (Berberich and Steiner, 2016). This may take a complex dimension when blockchain technology gets itself implemented on a regular basis for the technical ensuring of data security and nonrepudiation. In such a scenario, it may be a compelling reason to bring in a thorough change in the context of data protection laws so that there is a uniform law which can take care of protection of data even in the realm of a technological sphere present with or without blockchain technology.

The jurisdictional concerns are also capable of popping up in this realm of privacy and data protection because of involvement of persons from multiple jurisdictions whose activities are in turn intrinsically interconnected. *Per se* the content of blockchain itself is carrying information pertaining to the contributor/participants. The same itself is getting across borders and added with further blocks and is thus traversing multiple jurisdictional limits. Here, unlike other information in cyberspace, there is no difference between the so-called database and the controlling software. Definitely, a shared cloud-based information is also having more or less similar attributes. However, the unique position of blockchain is that the same is getting built up in a sequential manner and making it something which is incapable of an easy removal of intermittent data elements. In the process of this build-up, the contributors are capable of being present across multiple jurisdictions and data protection laws, like many other branches of law, are not uniform across nations. Owing to this aspect, data protection laws of specific countries are many a times demanding requirement of data protection laws on par with that of the originating country on being transferred to the receiving country.

The concepts of the originating country and the receiving country itself do not have much significance considering the nature of blockchain technology. There is difficulty in pinpointing to the originating country or the receiving country because right after the genesis block, it is more or less a linkage of blocks which in its technical essence is concerned only of the inherent techniques required in the build-up of blockchain. Any standard data protection law prescribes the option of taking back the consent. As mentioned earlier, the concept of consent is of much import in the context of data protection law. This itself was under challenge because of the nature of blockchain technology. Into this already complicated situation, another facet of data protection law is also stepping in, that is, the right of the person to take back the consent. In such a scenario, the data controllers are bound with the duty of removing such data elements. As mentioned earlier, this is technically not feasible by the nature of blockchain technology.

Altogether, with this aspect of data privacy and protection laws in the context of blockchain technology, it is discernible that the same is not painting a rosy picture. It is too difficult for the blockchain technology to survive by clinging on to the mandates of existing data protection law.

On the other hand, the need for norms with respect to data protection capable of even being used in the context of blockchain technology is still unavailable. Expecting a revisit to the existing data protection law so that the same can be utilized in the realm of blockchain technology is difficult to aspire for as many major jurisdictions itself are still debating a comprehensive data protection law and many a times relying on provisions spread out in different enactments under various subjects so as to give some solace for those who are claiming violations of privacy through data breach.

9.3.2 Contractual infirmities

The debate on the role of computers in the realm of contract making has been in existence for more than two decades (Waddams, 2017). It may not be wrong to say that the issues raised in the initial stage are still more or less continuing. The jural hangover in clinging to the fundamental aspects of contract formations is coming in the way of development of proper jurisprudence in the context of contract conclusion involving computers. Into this realm emerges the further complexities generated by blockchain technology.

The idea of smart contracts (McKinney et al., 2017; Werbach and Cornell, 2017; Temte, 2019a,b; Hsiao, 2017) is omnipresent in the technology realm (Raskin, 2016). However, how far the same can be legally acceptable without hampering the existing legal principles requires a serious discussion (Savelyev, 2017). The urge to fix the liability on a legal person is still thriving in the mindset of all those who are finding it difficult to appreciate the contract-making capability of software (Templin, 2019). The idea of contracts through blockchain technology is an interesting development when the same is closely monitored while reserving the expectations of the legal attributes.

In the context of blockchain technology, more than the idea of meeting of minds, it is the perpetuation of the involved set of blocks and the further contributors or participants. Each such contributor or participant is inherently becoming part of a larger arrangement that can be termed as a contract which by itself is indicating the inclination or interest of the parties. The idea of the presence of two parties and meeting of their minds, reciprocity of promise and acceptance, etc. are all incapable of being extended into the realm of blockchain technology (Lyons et al., 2019). In this context, the usage of conventional terms of contract might be a difficult task to identify the intent of the parties.

The law of contracts, right from its inception, has conceded to practical demands, and such an extension is the essential requirement in the context of blockchain as well. The idea of computers making contracts and bringing in newer concepts into the practical contractual requirements expected in the context of ICT needs to be read together, and the required concessions on the fundamental conventional attributes of law of contracts are to be made. The alteration to the law of contracts is a crucial aspect for bringing in a legal validity to the activities using or done through blockchain technology.

In the context of computers making contracts, the major concern is identifying a person who can be burdened with the contractual repercussions (Temte, 2019a,b). On the other hand, a computer artificially intelligent is capable of zeroing down on a contract in ways that are incapable of being apprehended or envisaged by the owner or creator of the said artificially intelligent system or software. In the realm of blockchain also, this aspect of contract conclusion is present and raising concerns. Here also, the ultimate concern is the legal compulsion of the presence of a responsible entity who can be made materialistically liable. This demand is, knowingly or unknowingly, relying on certain fundamental aspects, hanging in vacuum. In the context of evolving technology, it is only trying to encash the situation. The requirement is a total rethink on the entirety of the aspects taking into consideration the technological demand and the expected acceptance aspects of the stakeholders.

9.3.3 Property in and taxation of intangibles

The concept of property (Davies, 2007) in the realm of intangible kind is creating a value which is a fiction (Miller, 1910) upon a fiction. For example, gold as a valuable property is getting the so-called value because there is a set of persons valuing it. Nevertheless, the existence of the concept of gold is a reality. Coming to the realm of intangible goods, the so-called intangibility itself is indicating that there is vacuum and a creation of law. Rather, the so-called property is a fiction and the values are another fiction. Therefore, in the realm of intangible property, it all depends on the success of foretelling the fiction (Sherman, Bently, and Koskenniemi, 1999). Here, law is facing an irony. Hitherto the law is creating the fiction of property on one side, and the subjects ascribe the value related. More specifically, the state is capable of creating property in intangibles, and relying on the same, the subjects attribute or fictionize the value to the fiction of property without any leap and bounds.

Blockchain inherently possesses the attributes of security and nonrepudiation and is by itself creating a new type of property termed crypto asset (Allen et al., 2020). It is bringing a speculative platform for ascribing the values as per the freedom of the parties involved (Wójcik, 2021). This new type of property raises serious concern for the nation states in the context of jurisdiction. With the lack of necessity for law or state in the context of these crypto assets, there is a total absence of a legal system in that realm. The subjects of the earlier nation states are getting themselves liberated and becoming a state unto themselves. Thus, the fiction of property and the fiction of value are determined and fixed by the said closed self-sufficient system. This area is not merely a legal issue but an existential concern for the state itself. Crypto assets are many times getting referred to by the term cryptocurrency (Houben and Snyers, 2018). The term cryptocurrency is bringing in the idea that the same is something which is comparable to the legal tender of money. However, this attribute of legal tender of money is only present in the literal term currency (Perkins, 2018; He et al., 2016). Other than that, there is no comparison between the idea of legal tender of money and the terminology used for identifying crypto assets, that is, cryptocurrency. Irrespective of the terminology involved, the practical reality is that there are transactions of these crypto assets (Baker, 2015) and there is change of hands of conventional currency for these transactions. Rather, currencies are used for buying and selling cryptocurrencies. Definitely, this is an indication that there are crypto assets. As the term indicates, this is having all the attributes of intangible property. However, at the same time, there is a total absence of the usual player and nation state and definitely the absence of law as well. Two methods are possible in this context; one is recognizing such a type of crypto assets and trying to legalize the transactions involved in it, and the other is to ignore these phenomena and wait for the outcome. In this context, the issue of the concept of property attributes and its repercussions requires a serious consideration.

The above-mentioned paragraph gives an abstract level discussion of this scenario. Regulation of these transactions involving sale and purchase of crypto assets as well as the possibility to recognize the presence of such crypto assets is a topic of serious debate. This ultimately questions the taxation of the same.

The taxation of such a transaction is in a way recognizing the proprietary elements of crypto assets (Kochergin and Pokrovskaia, 2020). Therefore, in turn, it recognizes the fact that property itself is a fiction. On the other hand, the state is trying to tax these transactions, only looking at the transactional value ignoring the transaction. The attempts to tax such crypto assets are referring to the same and in a way recognizing its presence. Therefore, an argument of the absence of such a property is pointless. Even in the absence of a statutory regime to define and identify such a concept, there is legal acceptance of such an idea. This leads to the logical inference that the existing interest here is to enhance the revenue receipt for the transaction done.

Another realm of taxation influenced by blockchain technology is the scope of such a technology in narration of activities done by the subjects and making it beyond the scope identified by the state (PricewaterhouseCoopers, 2016). Blockchain technology is also referred to as distributed ledger. Through this, there is a scope for trading of anything of value securely and transparently. This is a sort of record, and hence, the terminology ledger is becoming appropriate. From a taxman perspective or that of a taxpayer complying with tax demands, such a record is becoming a necessary element. It generates an equal level of confidence in the data which are part of the ledger.

Blockchain is not a solution for all the ailments affecting the tax system but is capable of giving solutions to multiple concerns (Nemade et al., 2019). The current state of the tax system is facing multiple difficulties, and there are schools of thought that consider the need for a thorough revamping of the current tax system. In a strict sense, the emergence of blockchain is giving rise to scope for changes in the tax system so that it can withstand the present challenges. Blockchain as a technology can be a game changer in identifying tax incidents and methods of collecting the tax. In the context of blockchain, as mentioned earlier, the trust generation is happening between the interacting parties without the help or interference of a third party.

Blockchain technology usage to overcome the present challenges of the taxation system, at the same time, requires or challenges the established concepts of the taxation system at its definition level itself. The taxing incident will get recorded into the ledger irrespective of the conventional concepts relevant from the taxation perspective. The tax authorities will ultimately rely on these inputs into the distributed ledger for determining the expected taxation demands from the subjects. In such a scenario, there is a high possibility for the present ideals getting submerged and taking reincarnation in a new format conducive or rather permissive by the blockchain technology.

Anonymity is a factor omnipresent in ICT, wherein blockchain is an extended cut. There are concerns on possibilities of anonymous transactions using blockchain. Such a usage is a fertile ground for bringing in tax evasion. A series of transactions, either cyclical or taking a complicated route, are making no real-world implications. At the same time, creating a parallel sort of economic activity is an unavoidable possibility arising out of large-scale transactions involving multiple parties. The effectiveness of law in handling such situations is wait and watch. At the same time, as mentioned above, reliance on or mandating blockchain technology in the realm of transactions is a powerful tool to regulate the tendencies of tax avoidance. Thus, utilization of blockchain technology for transactions is a double-edged sword which can be used both ways.

9.4 CONCLUSION

Blockchain is only a technological development in its basic essence. Technological development, *per se*, is not a concern of law. However, when a technological development revolutionizes the existing societal interactions or opens up new methods for the same, the need for bringing in terms and conditions on such usage shall emerge. The larger the scope of new technological development, the wider are the challenges to subject areas of laws. When it is not finding a similar kind of concept, the challenges to law become more severe. Here, law is finding difficulties to identify an analogy and connect legal positions to rely so as to address the technological advancements. Blockchain technology is not an exception. It is capable of or has transformed substantial factors of societal movements. This demands a necessary legal interference to develop specific laws. As detailed, blockchain has been instrumental in influencing many facets of societal movements and thus has radically altered the existing patterns of interactions in the areas coming within the scope of cryptography, privacy and data protection, property laws, taxation, and contracts. In many of the fields, the legal system has identified the possibilities and the resultant challenges. The system has devised laws in a few areas among them. The future possibilities of blockchain in its further impact are only in the deliberation mode.

The international community and organizations have been continuing the role of a mute spectator rather than addressing the related apprehensions and legal concerns. However, the unilateral responses of the nation states hinder the further development of blockchain technology. Blockchain, as a technology, is a potential game changer. This reverberates policy and legal deconstruction of many of the existing structures and a rethink upon the utility of the same by analyzing the possibilities offered. The development of laws, thus in the context of blockchain, should be imbibed with this retrospection attitude and should also be in a homogeneous pattern owing to the borderless nature of ICT. Hence, there is a need for cooperation of nation states to bring forth laws for the respective jurisdictions and also to ensure that such laws are accommodative of each other.

REFERENCES

Allen, J. G., Rauchs, M., Blandin, A., & Bear, K. (2020). *Legal and Regulatory Considerations for Digital Assets*. https://papers.ssrn.com/sol3/papers.cfm?abstract_id=3712888

Baker, E. D. (2015). Trustless property systems and anarchy: How trustless transfer technology will shape the future of property exchange. *Southwestern Law Review, 45*, 351.

Bauer, F. L. (2002). *Decrypted Secrets: Methods and Maxims of Cryptology*. Springer Science & Business Media, pp. 2–8. https://link.springer.com/book/10.1007/978-3-540-48121-8

Berberich, M., & Steiner, M. (2016). Blockchain technology and the GDPR-how to reconcile privacy and distributed ledgers. *European Data Protection Law Review, 2*, 422.

Data Protection and Privacy Legislation Worldwide. https://unctad.org/page/data-protection-and-privacy-legislation-worldwide. (Last accessed on 12 February, 2022).

Davies, M. (2007). *Property: Meanings, Histories, Theories*. Routledge-Cavendish. https://www.routledge.com/Property-Meanings-Histories-Theories/Davies/p/book/9781904385844

De Filippi, P., Mannan, M., & Reijers, W. (2020). Blockchain as a confidence machine: The problem of trust & challenges of governance. *Technology in Society*, *62*, 101284.

Dong, Z., Luo, F., & Liang, G. (2018). Blockchain: A secure, decentralized, trusted cyber infrastructure solution for future energy systems. *Journal of Modern Power Systems and Clean Energy*, 6(5), 958–967.

Faden, R. R., & Beauchamp, T. L. (1986). *A History and Theory of Informed Consent*. Oxford University Press. https://global.oup.com/academic/product/a-history-and-theory-of-informed-consent-9780195036862?cc=in&lang=en&

Froomkin, A. M. (1994). Metaphor is the key: Cryptography, the clipper chip, and the constitution. *University of Pennsylvania Law Review*, *143*, 709.

Froomkin, A. M. (1996). The essential role of trusted third parties in electronic commerce. *Oregon Law Review*, *75*, 49.

Gardner, J. A. (1962). The sociological jurisprudence of Roscoe Pound (part I). *Villanova Law Review*, *7*, 1.

He, D., Habermeier, K. F., Leckow, R. B., Haksar, V., Almeida, Y., Kashima, M., et al. (2016). *Virtual Currencies and Beyond: Initial Considerations*. International Monetary Fund. https://www.imf.org/external/pubs/ft/sdn/2016/sdn1603.pdf

Hildebrandt, M. (2013). Extraterritorial jurisdiction to enforce in cyberspace? Bodin, Schmitt, Grotius in cyberspace. *University of Toronto Law Journal*, *63*(2), 196–224.

Houben, R., & Snyers, A. (2018). *Cryptocurrencies and Blockchain: Legal Context and Implications for Financial Crime, Money Laundering and Tax Evasion 15–52*. https://www.europarl.europa.eu/cmsdata/150761/TAX3%20Study%20on%20cryptocurrencies%20and%20blockchain.pdf

Hsiao, J. I. (2017). Smart contract on the blockchain-paradigm shift for contract law. *US-China Law Review*, *14*, 688–691.

Idrees, S. M., Nowostawski, M., Jameel, R., & Mourya, A. K. (2021). Security aspects of blockchain technology intended for industrial applications. *Electronics*, *10*(8), 951.

Kahn, D. (1996). The Codebreakers: The Story of Secret Writing from Ancient Times to the Internet. Scribner.

Kaur, R., & Kaur, A. (2012). Digital signature. In *2012 International Conference on Computing Sciences* (pp. 295–301). IEEE.

Kochergin, D., & Pokrovskaia, N. (2020). International experience of taxation of crypto-assets. *HSE Economic Journal*, *24*(1), 53–84.

Kulhari, S. (2018). Data protection, privacy and identity: A complex triad. In *Building-Blocks of a Data Protection Revolution: The Uneasy Case for Blockchain Technology to Secure Privacy and Identity* (1st ed., pp. 23–37). Nomos Verlagsgesellschaft mbH.

Kulhari, S. (2018). The midas touch of blockchain: Leveraging it for data protection. In *Building-Blocks of a Data Protection Revolution: The Uneasy Case for Blockchain Technology to Secure Privacy and Identity* (1st ed., pp. 15–22). Nomos Verlagsgesellschaft mbH.

Lemley, M. A. (2003). Place and cyberspace. *California Law Review*, *91*(2), 521–542. https://doi.org/10.2307/3481337.

Lerch, I., & Gray, M. (1997). Cryptography in America. *Science*, *278*(5343), 1545–1545.

Lyons, T., Courcelas, L., & Timsit, K. (2019). Legal and regulatory framework of blockchains and smart contracts. *In the Thematic Report for The European Union Blockchain Observatory and Forum* (Vol. 27, pp. 23–32).

Markey, H. T. (1983). Jurisprudence or juriscience. *William & Mary Law Review*, 25, 525–527.

McKinney, S. A., Landy, R., & Wilka, R. (2017). Smart contracts, blockchain, and the next frontier of transactional law. *Washington Journal of Law, Technology & Arts*, 13, 313.

Menthe, D. C. (1997). Jurisdiction in cyberspace: A theory of international spaces. *Michigan Telecommunications & Technology Law Review*, 4, 69.

Miller, S. T. (1910). The reasons for some legal fictions. *Michigan Law Review*, 8(8), 625.

Moore, A. (2018). Privacy, interests, and inalienable rights. *Moral Philosophy and Politics*, 5(2), 327–355. https://doi.org/10.1515/mopp-2018-0016.

National Academies of Sciences, Engineering, and Medicine. (2018). *Decrypting the Encryption Debate: A Framework for Decision Makers*. Washington, DC: The National Academies Press. https://doi.org/10.17226/25010.

National Research Council. (1996). *Cryptography's Role in Securing the Information Society*. National Academies Press.

Nemade, A. E., Kadam, S. S., Choudhary, R. N., Fegade, S. S., & Agarwal, K. (2019). Blockchain technology used in taxation. In *2019 International Conference on Vision Towards Emerging Trends in Communication and Networking (ViTECoN)* (pp. 1–4). IEEE.

Paar, C., & Pelzl, J. (2009). *Understanding Cryptography: A Textbook for Students and Practitioners*. Springer Science & Business Media.

Perkins, D. W. (2018). *Cryptocurrency: The Economics of Money and Selected Policy Issues*. US Congressional Research Service, R45427.

PricewaterhouseCoopers, L. L. P. (2016). *How Blockchain Technology Could Improve the Tax System*. PwC.

Radlo, E. J. (1997). Legal issues in cryptography. *In the International Conference on Financial Cryptography* (pp. 259–286). Springer, Berlin, Heidelberg.

Raskin, M. (2016). The law and legality of smart contracts. *Georgetown Law Technology Review*, 1, 305.

Reyes, C. L. (2017). Cryptolaw for distributed ledger technologies: A jurispruden-tial framework. *Jurimetrics*, 58, 283, 285.

Salmon, J., & Myers, G. (2019). *Blockchain and Associated Legal Issues for Emerging Markets*.

Savelyev A. (2017). Contract law 2.0: 'Smart' contracts as the beginning of the end of classic contract law. *Information & Communications Technology Law*, 26(2), 116–134, DOI: 10.1080/13600834.2017.1301036.

Schwerin, S. (2018). Blockchain and privacy protection in the case of the European general data protection regulation (GDPR): A Delphi study. *The Journal of the British Blockchain Association*, 1(1), 3554.

Shearer, J., & Gutmann, P. (1996). Government, cryptography, and the right to privacy. *Journal of Universal Computer Science*, 2(3), 113.

Sherman, B., Bently, L., & Koskenniemi, M. (1999). *The Making of Modern Intellectual Property Law* (Vol. 1). Cambridge University Press.

Sprung, C. L., & Winick, B. J. (1989). Informed consent in theory and practice: Legal and medical perspectives on the informed consent doctrine and a proposed reconceptualization. *Critical Care Medicine, 17*(12), 1346–1354.

Templin, S. (2019). Blocked-chain: The application of the unauthorized practice of law to smart contracts. *Georgetown Journal of Legal Ethics, 32,* 957.

Temte, M. N. (2019a). Blockchain challenges traditional contract law: Just how smart are smart contracts. *Wyoming Law Review, 19,* 94–101.

Temte, M. N. (2019b). Blockchain challenges traditional contract law: Just how smart are smart contracts. *Wyoming Law Review, 19,* 102–110.

Trachtman, J. P. (1998). Cyberspace, sovereignty, jurisdiction, and modernism. *Indiana Journal of Global Legal Studies, 5*(2), 561–581. http://www.jstor.org/stable/25691120 (accessed on 17 February 2022).

Waddams, S. (2017). Contract Law and the Challenges of Computer Technology. *The Oxford Handbook of Law, Regulation and Technology,* 317.

Werbach, K. (2018). Trust, but verify: Why the blockchain needs the law. *Berkeley Technology Law Journal, 33*(2), 487–550.

Werbach, K., & Cornell, N. (2017). *Contracts Ex Machina. 67 Duke Law Journal 313.* Available at SSRN 2936294.

Wójcik, D. (2021). Financial geography II: The impacts of FinTech–Financial sector and centers, regulation and stability, inclusion and governance. *Progress in Human Geography, 45*(4), 878–889.

Yaga, D., Mell, P., Roby, N., & Scarfone, K. (2019). Blockchain technology overview. *arXiv preprint arXiv:1906.11078.*

Zapechnikov, S., Tolstoy, A., & Nagibin, S. (2015). History of cryptography in syllabus on information security training. In *IFIP World Conference on Information Security Education* (pp. 146–157). Springer, Cham.

Zhang, R., Xue, R., & Liu, L. (2019). Security and privacy on blockchain. *ACM Computing Surveys (CSUR), 52*(3), 1–34.

Chapter 10

Blockchain for securing the IoT-based smart wireless surveillance cameras

Aurangjeb Khan and Jayanthi M.
CMR University

CONTENTS

DOI: 10.1201/9781003282914-10

10.1 INTRODUCTION

Wireless smart surveillance cameras are the need of the smart city. It is an urgent requirement to minimize crime or any disaster and to improve discipline in society in many situations. Smart cities have been growing with the evolution of Internet of Things (IoT) as it has the power to bind the operational technology (OT) with information technology (IT), resulting in automation of the smart city machinery like smart traffic systems, smart streetlights, and so on. Security of the IoT devices is one of the concerns in rising cyberattack on the system by scanning IP addresses of the devices. Blockchain technology can be used to register all the camera nodes with its respective edge computing device as secured blocks in blockchain and will allow only the authorities to access the contents on demand from the distributed network of camera nodes. Cameras generate a huge volume of data. Storing and retrieving all these contents to/from the cloud directly may be a slow process as well as a waste of storage. This problem may be overcome through edge computing technology by adding each node under blockchain. This will be a good solution for a secured surveillance system in a smart city. Edge computing devices will filter the contents and will send the labeled pictures and videos to cloud storage, thus simplifying the search and saving a lot of time for authorities. Edge computing is also useful for taking quick action or alerting the situation automatically based on the input from any of the cameras in the smart system. This chapter discusses IoT-based smart wireless surveillance cameras, importance of blockchain for securing the system, issues and challenges in the implementation, existing systems, hardware and software requirement for the system, circuit diagram of the camera node, flowchart of the working system, architecture of the system with blockchain, design and implementation using blockchain technology, setting up Ethereum on the Raspberry Pi node, comparison of the model with and without blockchain, and limitations and future scope of the system.

10.1.1 Blockchain

Blockchain technology is commonly used in cryptocurrency business, and it has become more popular due to its security feature. It increases the system reliability and privacy because of decentralized network features. It is also known as distributed ledger technology [1]. Due to the reliability and security features of blockchain technology, which works on the concept of decentralized networks, it can be used with IoT projects like surveillance camera networks in smart cities to secure the sensitive data.

10.1.2 Ethereum

Ethereum is a technology which is also used in cryptocurrency like blockchain. It has many applications like digital money transfer, global payment systems, and many more. It is a programmable and distributed network of blockchain, which can help us to make our own node to verify blocks and all the data transactions. Ethereum is an open source that can be used for improving the security aspect [2]. Ethereum is a smart contract and multi-purpose platform designed to solve multiple tasks. Some of the applications of Ethereum are for

- Building decentralized applications – DApps.
- Providing a new set of trade-offs matching a variety of DApps.
- Increasing the deployment speed.
- Ensuring high security for all applications.
- Advancement and efficient interactions.

Ethereum has built an ultimate basic layer to provide the above solutions. It is a new blockchain with a native Turing-complete coding language that enables developers and programmers to write smart contracts for various decentralized applications. Using Ethereum, we can set the ownership rules, conditions, transactions modes, and the state functions [3].

10.1.3 Edge computing

Edge computing is also referred to as fog computing; it is mainly used in data reduction and converting data flow into the information that is ready for storing and processing at higher layers in the IoT architecture. Edge computing works like a local cloud where data get stored and processed, and immediate actions are performed based on the system requirement. An edge device can be a router or Raspberry Pi which takes the data from different sensors, cameras, and so on and then processes, filters, and sends the required data to the cloud for storage and for further processing.

10.1.4 Internet of things

IoT is the combination of OT and IT. OT is related to the hardware which may include various sensors, actuators, routers, and so on, and IT includes the Internet, the operating system, and different applications. With these features, an IoT project connects physical devices and everyday objects like fans, TV, refrigerators, smart speakers, biometrics, smart cameras, and so on to the Internet and has applications in monitoring and controlling the devices and making them automated. Some of the common examples of IoT implementations are smart parking, smart cameras, smart homes, smart traffic systems, and so on. Most of these IoT implementations are prefixed with smart; the main reason is that these IoT systems work on their own and take decisions based on the sensor data and the logic written for operating the actuators.

10.1.5 Cloud storage

Cloud storage is a service provided by many companies for hosting websites, storing data, email services, and many more. IoT cloud also includes data visualization features and many applications to control and monitor IoT devices and IoT-based data with security. Cloud storage physically spans with many servers, which sometimes may be at different locations, and the physical environment is generally owned and managed by the hosting company. The cloud storage providers are mainly responsible for keeping the data available, accessible, secured, protected, and running. We can pay for the space and get the storage for our website or application on the cloud; some cloud storages are free with limited features where we can register and use the services [4]. Almost in all the IoT projects, cloud storage is used to store and analyze the data. In this system as well, we will be using cloud storage for storing labeled images/videos for the references filtering from the edge computing devices.

10.2 IoT-BASED SMART WIRELESS SURVEILLANCE CAMERAS

IoT-based smart wireless surveillance cameras are the solution for a smart city for surveillance and securing the city and at the same time for maintaining the discipline in the city. Installing wireless cameras can reduce the wiring, avoid the disconnection, and can reduce the cost of implementation. All the camera nodes (camera with Raspberry Pi) can be wirelessly connected to the local server and the cloud, and all the edge devices (Raspberry Pi) can process and send the images/videos to the cloud for storage and for further processing. Being an IoT-based smart camera, all the camera nodes can filter the videos and trigger the messages to the control room for the required action and can instruct the actuators for any immediate action, and at the same time, it can send the labeled images/videos to the cloud for storage and further processing. Whenever we connect any wireless or wired devices through an IP address to the server, there is always a security threat as the IP address can be compromised or hacked, resulting in the misuse of the system. To protect our surveillance camera nodes from such attacks, we can use blockchain as the second layer security over the first layer password-protected network of cameras.

Arduino Uno (ref-Table 10.1) does not have a built-in Wi-Fi module, but it is very cost effective compared to Raspberry Pi. Therefore, if we plan to implement this system using Arduino, we can connect Arduino to the ESP 2.4 GHz Wi-Fi module, which can cover the area around 92 m, which is enough to push the data to the cloud and to the local server. Therefore, our camera node will consist of Arduino Uno, a camera, an ESP module, a battery, and a buzzer, whereas Raspberry Pi has a built-in Wi-Fi module and so

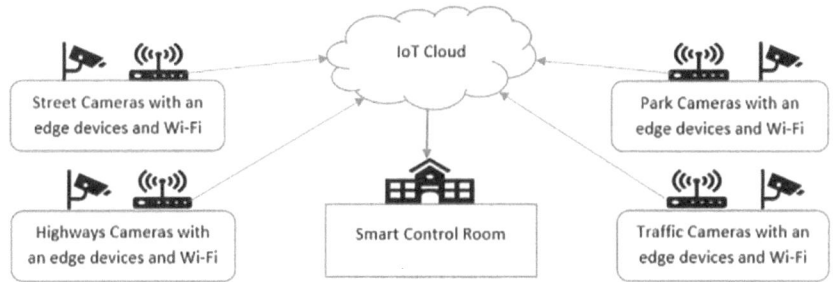

Figure 10.1 Smart cameras in a smart city with IoT cloud and a smart control room.

all the cameras can also be connected separately with Raspberry Pi, which acts as an edge device and can filter and process the data at the edge itself to make processing faster; this concept is explained in Figure 10.1. Once the camera settings are done with the edge devices, now for the robust security purpose, we can add all the edge devices to the blockchain to secure our surveillance cameras from intruders and can protect our edge devices from being misused. All the cameras with edge computing devices can be treated as a single block in the blockchain. Therefore, in a smart city, we will be having multiple nodes of cameras and all these nodes will be known as blocks of blockchain.

A lot of smart wireless cameras are available in the market, which can be easily installed at home and can be viewed using a smart phone; hence, it may be a good option for a home surveillance, but it lacks the edge computing and cannot perform the edge level processing which is required in a smart city for different situations; therefore, it will not be an optimal solution for a smart city for monitoring the streets, circles, parking, and roadways.

10.3 IMPORTANCE OF BLOCKCHAIN FOR SECURING THE SYSTEM

Blockchain is always secure due to its encryption feature and its low-cost transaction because of the decentralization feature of blockchain and the fact that the authenticity of the data transaction is verified by the participants; in this model, the participants are the control rooms of the smart city. Blockchain technology has been a verified and trusted technology in the field of cryptocurrency like Bitcoins, Ethereum, and so on, which are mostly trusted because of the security layer in the blockchain framework.

Blockchain technology is not limited to just the cryptocurrency, but its applications are more in the real-world business as listed below in the various areas:

- Sharing of medical data securely
- Music royalty tracking system
- Payment system across the border
- Real-time IoT operating system
- Personal identity security
- Anti-money laundering and tracking system
- Supply chain and logistics monitoring system
- Voting mechanism
- Advertising insights
- Original content creation
- Real estate processing platforms

Applications of Bitcoin have been still growing, especially in the IoT-based system. Our effort here through this chapter is to discuss the model of smart wireless surveillance cameras in the smart city with blockchain to secure the smart city cameras and to make it robust and really smart with the help of edge computing [5].

10.4 ISSUES AND CHALLENGES IN THE IMPLEMENTATION

There are many issues and challenges in the implementation of the smart surveillance cameras with respect to the implementation cost, image processing, and blockchain implementation along with edge computing devices.

- Glare images – It is difficult to process moving images in the running traffic, foggy weather images, night images due to low light, and sometimes images due to bad weather.
- Shadow of images – confuses the model to predict the actual situation; it may be difficult to predict the size, shape, and motion in shadow.
- Crowded images – It may be difficult to identify suspicious activity in crowded places; especially, theft detection and other crimes are challenging to figure out from the overcrowded places.
- Implementation of blockchain or Ethereum with edge computing nodes on camera networks requires additional memory to install the Geth/Ethereum node at each edge device, resulting in the increase of implementation cost.

10.5 EXISTING SYSTEM

In most of the cities, CCTV cameras are installed at multiple locations, which are mostly connected physically to the digital video recorder (DVR), where videos get recorded constantly in the DVR, and if there is a need to do some investigation, the authority has to view the complete recording of the day, which is very time-consuming. In some other cases, CCTVs are connected physically or using the wireless mode to the control rooms; these cameras record videos constantly in the DVR at the control rooms, where authorities can also watch these videos live and can take any action required. In some cases, the camera input is programmed with a traffic application to take necessary action for signal jumping, breaking traffic rules, and so on.

There are four problems in the existing system; the first problem is that most of these cameras are not powered by edge computing and due to the huge volumes of videos are sent to the control room/s, which is a very time-consuming process to check the particular videos. The second problem is that it is not a smart camera because of no edge device; hence, it does not send alert signals to the control room if any accident happens or if any unusual incident is captured by the camera. The third problem is that it does not differentiate the normal and unusual images or videos, so it does not send any labeled images to the control room to take immediate action. The fourth problem is that it does not trigger any actuator or does not take any action automatically if any incident is captured.

10.6 HARDWARE AND SOFTWARE REQUIREMENT FOR THE SYSTEM

We are discussing two different sets of hardware and software required to implement the smart cameras with edge computing devices in a smart city, one using Raspberry Pi on Raspbian OS, where coding can be done using python programming, and the second using Arduino Uno with ESP 2.4 GHz Wi-Fi modules. For Arduino, coding will be done using C++ programming. Arduino Uno is much cheaper than Raspberry Pi, but it does not have built-in Wi-Fi, so we need to connect cameras with the ESP Wi-Fi module.

10.6.1 Hardware and software requirement for the smart camera system using Raspberry Pi

Raspberry Pi hardware set:

- Raspberry Pi
- Camera

- Rechargeable battery
- Relay to connect the camera with Raspberry Pi
- Buzzer to make noise when required
- Case to cover the board, camera, battery, and relay in a single unit

Software:

- Raspbian Operating System
- Python programming and its libraries
- IoT cloud

10.6.2 Hardware and software requirement for the smart camera system using Arduino Uno

Arduino Uno hardware set:

- Arduino Uno board
- ESP Wi-Fi module
- Camera
- Rechargeable battery
- Relay to connect the camera with the ESP module and Arduino Uno
- Buzzer to make noise when required
- Case to cover the board, camera, battery, and relay in a single unit

Software:

- Arduino Software (IDE)
- C++ Programming on IDE
- IoT Cloud

Dedicated Wi-Fi Internet connection is required for all the edge devices and to integrate all the camera nodes (cameras and Raspberry Pi/Arduino Uno/ESP) with control rooms through IoT cloud. We would like to use a buzzer/speaker along with cameras to make noise to alert any specific action in the place of the event. Edge devices will trigger this buzzer to alert based on the specific analysis; at the same time, the edge device will send a mail/message to the control room to take any immediate action.

Let us check the images of the main components of the surveillance system in Table 10.1.

Table 10.1 Components of the surveillance camera system

Raspberry Pi: This is a debit card size minicomputer with multi-port like the following: USB, Ethernet, HDMI port, micro-USB for power, audio port, Camera port, Display port, micro-SD card port for storage, 40 GPIO pins, Bluetooth 4.1, Wi-Fi, CPU, GPU, and memory. It runs on Raspbian OS [6].

Cameras: A lot of different types of cameras are available with USB and CSI. We can prefer to use a USB-based camera because it comes with good power and night vision. For highways and streets, a powerful camera must be used for better clarity.

Relay: This is used to connect a security camera which uses more than 5V and a buzzer/horn with Raspberry Pi to supply sufficient power from another socket.

Arduino Uno: This is a simple microcontroller motherboard which runs a single code over and over again; if the second code is uploaded to the board, the first one gets replaced, and it does not run on OS like Raspberry Pi.

(Continued)

Table 10.1 (Continued) Components of the surveillance camera system

ESP Wi-fi Module: Arduino Uno does not have built-in Wi-Fi, so if we are going to build this system on it, we need to use the ESP Wi-Fi module to send signals on the Internet.

Buzzer: will be used to make noise to prevent street fights or alert passersby if any incident happens in the public place. It will ring automatically based on specific videos captured on the camera.

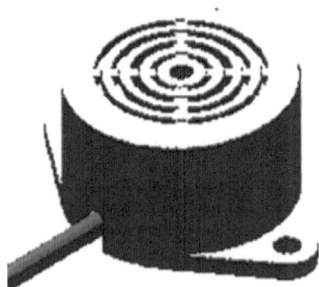

Battery: We need a rechargeable battery to power the devices (Raspberry Pi, camera, buzzer) 24×7.

10.7 CIRCUIT DIAGRAM OF A CAMERA NODE

Most surveillance cameras for Raspberry Pi come with a USB or ribbon connection which can be connected directly to Raspberry Pi without any external power supply, but it is not efficient for the implementation on the roadways or street as it must be more powerful and must be efficient in night vision as well. A powerful camera for such a scenario mostly works on 12–50 V [7], and as we know, Raspberry Pi can supply a maximum of 5 V; thus, to power a powerful camera, we need to connect it using a relay as shown in Figure 10.2, and the relay can be controlled from Raspberry Pi. Similarly, a buzzer can be given external power so that it can produce sufficient noise and can be controlled from Raspberry Pi as per the condition.

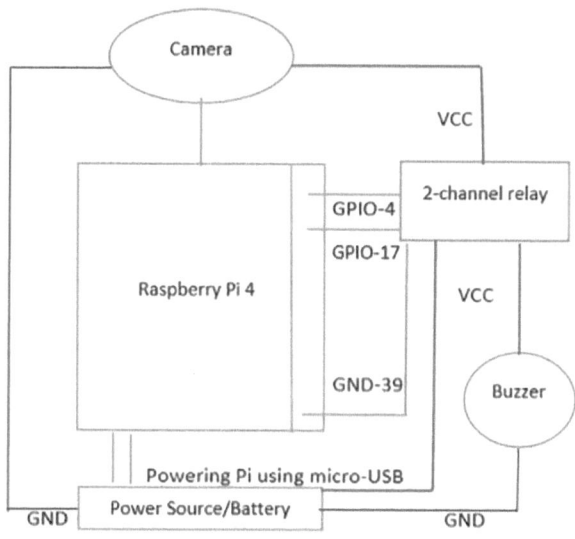

Figure 10.2 Camera node connection with Raspberry Pi, a relay, a camera, a buzzer, and a power bank.

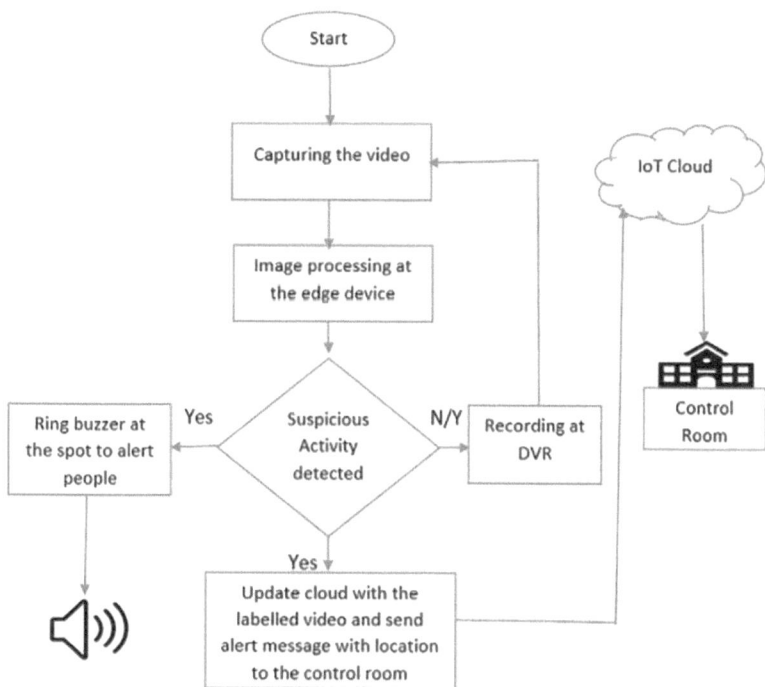

Figure 10.3 Flowchart representing the working of the smart surveillance camera.

10.8 FLOWCHART OF THE WORKING SYSTEM

Let us understand the smart surveillance camera system with the help of the flow chart, which includes capturing the video, processing the video at the edge computing device using an image processing algorithm, and taking appropriate action based on the suspicious activity detection (Figure 10.3).

10.9 ARCHITECTURE OF THE SYSTEM WITH BLOCKCHAIN

In this section, we discuss the detailed architecture of the smart surveillance system with the blockchain framework. Before the detailed architecture, we will make a node of the smart camera (Figure 10.4).

10.9.1 Architecture of the object detection by a smart surveillance system using different algorithms

Implementation of object detections required a lot of image processing algorithms to be implemented at the edge computing device; many researchers have proposed different techniques for the object detections and analysis, which are listed in this architecture and are cited [8,9] (Figure 10.5).

Static Object Detection: This module is to detect static objects like abandoned bags/boxes to prevent the explosive attack in the city. In the video analysis of the camera, the video's backgrounds are considered as moving objects like foreground objects and the static objects are

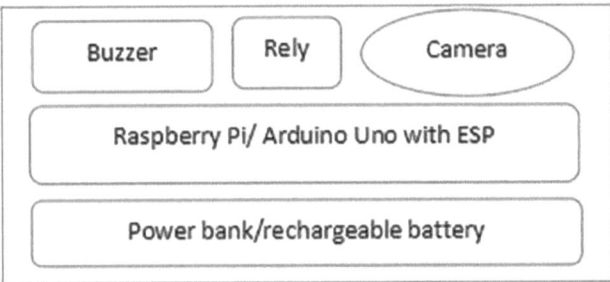

Figure 10.4 Smart cam node containing the Pi board, a camera, a buzzer, a relay, and a power bank.

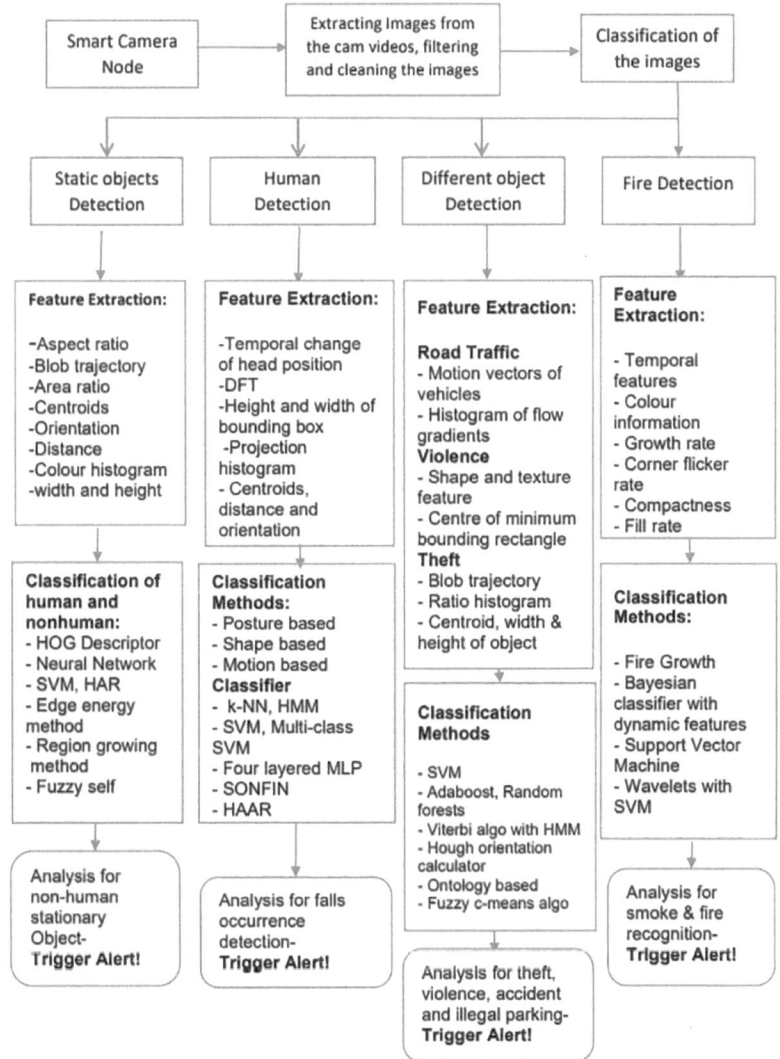

Figure 10.5 Object detection architecture based on different algorithms.

considered as background objects. Therefore, when a newly detected object becomes still or static, then it is considered as absorbed in the background. Many authors have used different background subtraction techniques with a dual background approach, with different methods to extract the two foreground objects for detecting the abandoned object in the video. Porikli and team [10] have proposed a video surveillance system which uses dual foreground extraction from dual background modeling techniques.

Human Detection: This module is specially used to detect and analyze the human fall based on the shape, posture, and motion classification; to detect and analyze this occurrence, we need to implement multiple algorithms, which are proposed by different researchers and are cited here: Auvinet and team [11] have presented the fall detection based on 3-D silhouette vertical distribution, Liu [12] has presented the improved algorithm for detecting fall, Foroughi and team [13] have presented the technique to detect fall based on the multi-class support vector machine and human shape, and Thomas and team [14] have presented real-time fall detection based on the analysis of posture estimation.

Different Object Detection: Research has been done on the moving foreground object detection from the videos of surveillance systems. This technique helps us in extracting the human activities such as attacks, chain snatching, fighting, running, crowds, vandalism, hitting, and falling in surveillance videos. Some of these approaches/techniques for implementation are cited here for this module: Wren and team [15] have presented an independent background modeling method at each pixel location using a single Gaussian for the real-time tracking of the human body. Stauffer and Grimson [16] have proposed a most common background model based on the mixture of Gaussians, but this technique cannot model accurately to the fast variations of the background. Elgammal and team [17] have developed a non-parametric model to model a background which is based on Kemel density estimation (KDE); it is good for fast variations in the background as well. Lo and Velastin [18] have proposed the temporal median filter background technique. Cucchiara and team [19] have proposed an approach for moving object and shadow analysis. Piccardi [20] has presented a review on seven different methods. Bouwmans [21] has provided a complete survey of traditional and recent background modeling techniques to detect the foreground objects from the static camera's video.

Fire Detection: Many researchers have proposed the techniques on fire and smoke detections; some of these works are cited here: Chen and team [22] have published the work to detect smoke and fire based on image processing. Toreyin and team [23] have presented the real-time fire and flame detection. Foggia and team [24] have also presented the real-time fire detection for video surveillance, which is very helpful for this model to implement.

10.9.2 Architecture of the surveillance system with blockchain

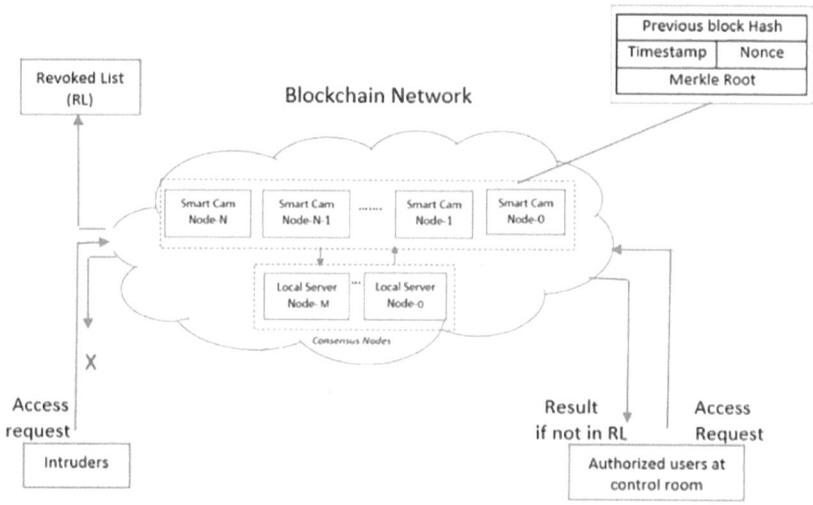

Previous block Hash	
Timestamp	Nonce
Merkle Root	

Figure 10.6 Surveillance system architecture with blockchain; each smart camera node is an independent node with an edge computing power; the system must be registered on the blockchain to ensure the node's security.

10.10 DESIGN AND IMPLEMENTATION USING BLOCKCHAIN TECHNOLOGY

10.10.1 System design

Surveillance camera systems using blockchain will ensure three-layer security to the cam network gateways; one uses the group signature scheme, the second message authentication code is for cam gateways, and the third is MAC with a chosen key pair, which ensure only the authorized access to the cam gateways as shown in Figure 10.6. The four modules that can be focused are as follows:

System Setup: In this module group, public key and group private key setup has to be done. Authorized members are allocated with a group private key for communicating with the node if needed and a log file to be updated for every access by the members.

Request Control: For control room members who wish to get access to the cam gateways, a new public/private key pair (pk, sk) is generated. This is one of the preferred approaches to avoid unauthorized access to the cam node.

State Delivery: The cam gateway receives new requests through the smart contract module. Once the control room member requests to access the cam node, the request is retrieved by the cam gateway through the get_request method. If the transactions get verified by the G-Verify method and have not been added in the revocation list (revoList), then the cam gateway decrypts by the **Dec** method using its private key. Basically, this module is to verify the access.

Chain Transaction: This module is for monitoring the transaction through smart contract and to discard the unauthorized accesses which are added in the revoList.

10.10.2 Implementation with blockchain

For the implementation of the smart surveillance system of a smart city, we propose the technology which is already known for its robust security characteristics, that is, blockchain. Here, we are proposing the smart surveillance system using the Practical Byzantine Fault Tolerance (PBFT) algorithm [25]. This algorithm allows us to achieve thousands of transactions per second, so any camera node can transfer the images to the control room, and at the same time, any camera can be accessed by the authorized user. Local server nodes could be some servers with sufficient computational speed and storage at the control room.

10.10.3 Transactions in blockchain

In cryptocurrency, a transaction includes a signature, the address of the sender and receiver, and the transferred amount/data. In our proposed module, a request transaction includes a version, a fresh public key, device information, and control order, which is a message including the transaction version, public key, device information, and control order. This message needs to be encrypted and to be used by the control room gateway's public key and then signed by group signature using the user's group private key gsk. Suppose if a person wanted to access the particular camera node, he/she needs to obtain a fresh public/private key pair and then complete the transaction TX = GSign (Enc (01 ‖ pk ‖ cam(i) ‖ O), where 01 is the current version, cam(i) is the camera identity, O is the control order, Enc denotes the data encryption with the Elliptic Curve Integrated Encryption Scheme (ECIES), pk is the gateway public key, and gsk is the control room user's group private key [26,27].

10.10.4 Smart contracts in blockchain

Smart contracts are mainly programs that are stored on the blockchain and run when predetermined thresholds or conditions are matched. They are basically used to automate the execution of an algorithm/logic so that all valid users can be sure of the outcomes immediately, without any third party's involvement or time loss. It also automates the workflow by triggering the next action when certain conditions are matched [28]. Many programming languages (Solidity, Rust, JavaScript, Vyper, Yul, Python, and Go) are used to write smart contract codes. After compiling the smart contract, it is recorded into the chain as a piece of bytecode in the transaction. Methods of the smart contract can be triggered by a message call from another contract; other transactions are not generated by this message call.

Smart contracts can be used in the surveillance system to record the access request from the users and response from cam node gateways. According to algorithm #1, a user requests (containing the targeted cam gateway's public key pk and its group signature Sign) into the smart contract through the **uploadReq** function. After that, the request and responses are validated through the **uploadRes** function and will be monitored by the cam gateway. Finally, the request list will be deleted periodically using the **deleteReq** function by the control room to manage the storage system of the smart contract. At the same time, local servers at the control room will also maintain the valid requests into the blockchain for the future inspection. Smart contract is also used to manage the revocation list (RL). The consensus node/local server can easily identify the valid request based on RL; consensus nodes also verify the signature of the transaction as well as the authority of the issuer by accessing the RL. When the transaction signature is valid and the private key component does not make it to the RL, then it is treated as a valid transaction.

Operations of RL include uploading, deleting, and reading, which are demonstrated here. **getReq, getRes,** and **getRL** are the view types of functions.

```
Algorithm 10.1: Smart contract for smart surveillance
cameras
Require: calling the function with parameter
Ensure: Setting up functions with the properties
#Request structure
Structure Request:
Unsigned int128[ ] Sign              # Sign-Group signature.
Unsigned int128[ ] Result
def SmartSurveillance():
# Function that gets triggered automatically at the run time
Manager = msg.sender
mapping (addr => Request) public reqList
Unsigned int128 [ ] revoList
def uploadReq(addr, groupSign):
# It is triggered when a user requests
reqList[addr].Sign = groupSign
def uploadRes(addr, state):
# It is triggered to respond to a task by the cam gateway
        require(msg.sender == addr)
```

```
                    reqList[addr].Result = state
def view getReq(addr):
# It is triggered to obtain the request by the cam gateway
require(msg.sender == addr)
        return reqList[addr].Sign
def getRes(addr):    # a view function
# It is triggered to get a request by the authorized user
return reqList[addr].result
def deleteReq(addr):
# It is triggered to clear RL to manage the memory of the
device
require(msg.sender == Manager)
        reqList[addr].Sign = NULL
        reqList[addr].Result = NULL
def addRL(privateInformation):
# It is triggered to add authorized access to the RL.
require(msg.sender == Manager)
        if revoList.isNotExist(privateInformation):
        revoList.push(privateInformation)
        return 1;
        else:
        return 0;
def deleteRL(privateInformation):
# It is triggered to delete the RL list
require(msg.sender == Manager)
        r = revoList.find(privateInformation)
        free(revoList[r])
def getRL():                         # a view to get the list of RL
# It is triggered to get the list of RL from the control room
return revoList
```

10.11 SETTING UP ETHEREUM ON THE RASPBERRY PI NODE

Ethereum is a programmable blockchain and a distributed network of computers which can help us to make our own node and to verify blocks and all transaction data. Once our camera node is set up with the Raspbian operating system, we can set up Ethereum on Pi to make our camera node as the blockchain node powered by Ethereum.

Steps to install Ethereum on the Pi node: [29]

Step 1: Installing Ubuntu on the Pi – we can select Ubuntu from the imager and write on the SD card, the same way as the Raspbian installation on the SD card.

Step 2: After Ubuntu on Pi, we will be able to access Pi remotely; now, we can update the OS using the **update and upgrade** commands.

Step 3: Installing Unzip using the command $ Sudo apt-get install unzip

Step 4: To install go language using the command $ Sudo snap install go –classic

Step 5: Installing make using the command $ Sudo apt-get install make: It is used for compiling and managing a collection of applications or files from the source code.

Step 6: Installing htop using the command $ Sudo apt install htop: It allows us to monitor the systems and its resources or the server processes in real time; installing build-essentials: $ Sudo apt install build-essential

Step 7: Build-essentials does not install anything but provide the links to several other packages that will be installed as dependencies; this is the reason it is called a meta-package.

Step 8: Assigning a static IP address to Pi, editing the netplan file using the text editor, by default, Pi will be having a dynamic IP address. We can apply the changes on netplan using the command

$ Sudo netplan apply, #The system needs to be rebooted

Step 9: Mounting HDD/SSD for storing the state of the blockchain on Pi because it requires sufficient memory resources; we can connect the storage using USB of Pi and can mount using the following commands:

$ Sudo fdisk -l ## for listing available disks

$ sudo mkfs.ext4 /dev/sda ## for creating a partition on the disc

$ sudo mkdir /mnt/ssd ##for mounting the disc

$ sudo chown -R ubuntu:ubuntu /mnt/ssd/

$ sudo mount /dev/sda /mnt/ssd ## mounting the disk "dev/sda" to "/mnt/ssd"

Step 10: mounting the disc automatically on startup of Pi

First, we need to get the unique id of the disc [UID = "<the unique id>"]

$ Sudo blkid

$ Sudo vim /etc/fstab ## for editing/inserting UID at the end of the /fstab file.

$ Sudo reboot ## to reboot the system after editing the file.

$ df -ha /dev/sda ## to check if the disc is mounted.

Step 11: Now we should add swap space so that it can help mitigate the memory problem, especially when the RAM is full due to the operations using Geth.

Step 12: Installing core-geth

$ git clone https://github.com/etclabscore/core-geth.git && cd core-geth/

Clone the repo and change directory into source

$ make geth ## command to make geth

$ Sudo mv ~/core-geth/build/bin/geth /bin/ ##Moving the geth/ built to the /bin/folder

$ geth version ## To check the installation

$ geth –help ## To view the geth usages and commands

Step 13: Running core-geth:

Geth stores data on ~/ethereum/geth on the micro-SD by default; we can change its directory

$ Sudo mkdir /mnt/ssd/ethereum/ ## creating Ethereum directory

$ Geth --syncmode fast --cache 256--datadir /mnt/ssd/ethereum ##running geth

$ nohup geth --classic --syncmode fast --cache 256--datadir / mnt/ssd/classic ## to run in the background; otherwise, Geth runs in the foreground by default.

$ htop ## to check the geth running status

$ Sudo geth attach ipc:/mnt/ssd/classic/geth.ipc ## To enter into the Geth console >eth.syncing ## to get the syncing status of Ethereum

10.12 COMPARISON OF THE MODEL WITH AND WITHOUT BLOCKCHAIN

The surveillance system of a smart city with blockchain, as we discussed in the above sections, required access verification using a public key and the group signature to differentiate between intruders and the authorized users, and in this system, if any unauthorized user tries to access the cam node, his/her request will be denied and will be pushed into the RL; the user which is not in the RL will only get the access to the node; hence, with the properties of the blockchain framework, the system is robust and completely secured. Each node of the surveillance system will run the Ethereum to ensure the security as explained in Section 10.11, but at the same time, the system implementation cost increases due to the extra memory requirement to store the state of the blockchain at each node. However, in the surveillance system without blockchain, all this verification cannot be done, and for security of the cam nodes, we need to depend on the traditional security system only, that is, password; hence, there are chances of attack on the nodes as the hacker can easily scan the IP address of the nodes getting into the network. Moreover, any of this system whether with or without blockchain requires a lot of image processing algorithms (as mentioned above in Section 10.9.1) to be implemented to analyze the different objects and activities based on the threshold triggering the message and actuator.

10.13 LIMITATION AND FUTURE SCOPE

The implementation cost of the cam node with blockchain is going to be high as the node may require additional memory for storing the blockchain state/

installing the Geth, and the implementation of various image processing algorithms for all the aspects on the Pi node is also a challenging task. We have a good list of image processing algorithms (Section 10.9.1) for analyzing the different activities after taking images from the surveillance videos. The study of the comparison between these algorithms could be done. In future, we will look forward to doing the comparison of all these algorithms category-wise and do the implementation with best-suited algorithms from each category.

10.14 CONCLUSION

In this chapter, we have proposed a model for a surveillance system using IoT for a smart city with a blockchain framework to make it completely secure. We have stressed on using the edge computing device at each camera node for implementing the image processing algorithms to do the classification of different types of activities and pushing messages to the control room and triggering buzzer based on the threshold. The workflow architecture is presented for the classification of different objects and activities using different sets of image processing algorithms. The design and implementation section highlights about the implementation of blockchain technology for securing the camera nodes from intruders.

REFERENCES

1. Blockchain Council. (n.d.). *What Is Blockchain Technology, and How Does It Work?* Retrieved Dec 22, 2021 from https://www.blockchain-council.org/blockchain/what-is-blockchain-technology-and-how-does-it-work/.
2. Ethereum. (n.d.). *Welcome to Ethereum.* Retrieved Jan 3, 2022 from https://ethereum.org/en/.
3. Applicature. (2018). *Ethereum Programming Tutorial.* https://applicature.com/blog/blockchain-technology/ethereum-programming-tutorial.
4. Wikipedia. (n.d.). *Cloud Storage.* Retrieved Jan 5, 2022 from https://en.wikipedia.org/wiki/Cloud_storage.
5. Daley, S. (2021). *34 Blockchain Applications and Real-World Use Cases Disrupting the Status Quo.* Retrieved from https://builtin.com/blockchain/blockchain-applications.
6. Raspberry Pi. (n.d.). *What Is a Raspberry Pi?* Retrieved Dec 23, 2021 from https://www.raspberrypi.com/products/raspberry-pi-3-model-a-plus/.
7. Caputo, A. C. (2014). *3-Digital Video Hardware. Digital Video Surveillance and Security* (second ed.), Butterworth-Heinemann, Boston.
8. Tripathi, R. K., Jalal, A. S., & Agrawal, S. C. (2018). Suspicious human activity recognition: A review. *Artificial Intelligence Review*, 50(2), 283–339.
9. McHugh, J. M., Konrad, J., Saligrama, V., & Jodoin, P. M. (2009). Foreground-adaptive background subtraction. *IEEE Signal Processing Letters*, 16(5), 390–393.

10. Porikli, F., Ivanov, Y., & Haga, T. (2007). Robust abandoned object detection using dual foregrounds. *EURASIP Journal on Advances in Signal Processing*, 2008, 1–11.

11. Auvinet, E., Multon, F., Saint-Arnaud, A., Rousseau, J., & Meunier, J. (2010). Fall detection with multiple cameras: An occlusion-resistant method based on 3-d silhouette vertical distribution. *IEEE Transactions on Information Technology in Biomedicine*, 15(2), 290–300.

12. Liu, H., & Zuo, C. (2012). An improved algorithm of automatic fall detection. *AASRI Procedia*, 1, 353–358.

13. Foroughi, H., Rezvanian, A., & Paziraee, A. (2008). Robust fall detection using human shape and multi-class support vector machine. In *2008 Sixth Indian Conference on Computer Vision, Graphics & Image Processing* (pp. 413–420). IEEE.

14. Thome, N., Miguet, S., & Ambellouis, S. (2008). A real-time, multiview fall detection system: A LHMM-based approach. *IEEE Transactions on Circuits and Systems for Video Technology*, 18(11), 1522–1532.

15. Wren, C. R., Azarbayejani, A., Darrell, T., & Pentland, A. P. (1997). Pfinder: Real-time tracking of the human body. *IEEE Transactions on Pattern Analysis and Machine Intelligence*, 19(7), 780–785.

16. Stauffer, C., & Grimson, W. E. L. (1999). Adaptive background mixture models for real-time tracking. In *Proceedings 1999 IEEE Computer Society Conference on Computer Vision and Pattern Recognition (Cat. No PR00149)* (Vol. 2, pp. 246–252). IEEE.

17. Elgammal, A., Harwood, D., & Davis, L. (2000). Non-parametric model for background subtraction. In *European Conference on Computer Vision* (pp. 751–767). Springer, Berlin, Heidelberg.

18. Lo, B. P. L., & Velastin, S. A. (2001). Automatic congestion detection system for underground platforms. In *Proceedings of 2001 International Symposium on Intelligent Multimedia, Video and Speech Processing. ISIMP 2001 (IEEE Cat. No. 01EX489)* (pp. 158–161). IEEE.

19. Cucchiara, R., Grana, C., Piccardi, M., & Prati, A. (2003). Detecting moving objects, ghosts, and shadows in video streams. *IEEE Transactions on Pattern Analysis and Machine Intelligence*, 25(10), 1337–1342.

20. Piccardi, M. (2004). Background subtraction techniques: a review. In *2004 IEEE International Conference on Systems, Man and Cybernetics (IEEE Cat. No. 04CH37583)* (Vol. 4, pp. 3099–3104). IEEE.

21. Bouwmans, T. (2014). Traditional and recent approaches in background modeling for foreground detection: An overview. *Computer Science Review*, 11, 31–66.

22. Chen, T. H., Wu, P. H., & Chiou, Y. C. (2004). An early fire-detection method based on image processing. In *2004 International Conference on Image Processing, 2004. ICIP'04.* (Vol. 3, pp. 1707–1710). IEEE.

23. Töreyin, B. U., Dedeoğlu, Y., Güdükbay, U., & Cetin, A. E. (2006). Computer vision based method for real-time fire and flame detection. *Pattern Recognition Letters*, 27(1), 49–58.

24. Foggia, P., Saggese, A., & Vento, M. (2015). Real-time fire detection for video-surveillance applications using a combination of experts based on color, shape, and motion. *IEEE Transactions on Circuits and Systems for Video Technology*, 25(9), 1545–1556.

25. Castro, M. & Liskov, B. (2002). *Practical Byzantine Fault Tolerance and Proactive Recovery.* Retrieved from https://doi.org/10.1145/571637.571640.

26. Lin, C., He, D., Kumar, N., Huang, X., Vijayakumar, P., & Choo, K. K. R. (2019). Homechain: A blockchain-based secure mutual authentication system for smart homes. *IEEE Internet of Things Journal*, 7(2), 818–829.

27. Hankerson, D., Vanstone, S., Menezes, A. (2004). *Guide to Elliptic Curve Cryptography*, vol. 22, no. 03, (pp. 189–311).

28. IBM. (n.d.). *What Are Smart Contracts on Blockchain?* Retrieved Jan 4, 2022 from https://www.ibm.com/topics/smart-contracts.

29. Core-Geth. (n.d.). *How to Setup an Ethereum Node on Raspberry Pi.* Retrieved Jan 28, 2022 from https://core-geth.org/setup-on-raspberry-pi.

## Chapter 11

Central bank digital currency

Vidyashankar Ramalingam
Independent Blockchain Consultant

CONTENTS

11.1 INTRODUCTION

Money as a form of exchange is an IOU (promise to pay) that performs an important role in the society; it acts as a unit of account, a medium of exchange and a store of value. The physical form of fiat currency that we use currently has taken a long road to reach the place where it currently stands, from being an alternative to the barter system and to get evolved into multiple forms like coins, currency notes and cryptocurrencies.

The current form of currency held by the public is built on a solid financial structure wherein the central bank and the commercial banks play an important role in maintaining this eco-system.

At present, banks and financial institutions alone can hold the electronic form of central bank money as reserves. Whereas the public can hold only the bank notes issued by the central bank and with the advent of the cryptocurrencies and stable coins, it is imperative for central banks to stay relevant with the innovation. The central bank money issued electronically is called the central bank digital currency (CBDC), which would facilitate business, financial institutions and households to make transactions seamlessly in a robust manner and enable storage of value, thus promoting the innovation with multiple use cases built on top of CBDC by enabling opportunities and access to the central bank money.

This chapter explores CBDC and its operational structure with respect to different types, design features to be considered while designing a CBDC model and potential use cases where the CBDC could be considered beneficial. CBDC has gained prominence in the recent years post the success of blockchain technology in successfully creating a decentralized trust and eco-system for business use cases. This chapter explains how blockchain technology could be used in the implementation of CBDC, live examples of how different jurisdictions have explored CBDC and risks of not implementing the same.

This chapter consists of the following key sections:

- **Central Bank Digital Currency – Introduction:** This section outlines the basic definition and the functions of CBDC.
- **Design Choices:** This section discusses the design choices to be considered when implementing CBDC.
- **Application of Distributed Ledger Technology:** This section explains how blockchain technology could be leveraged for implementing CBDC.
- Global Projects on CBDC

11.2 CENTRAL BANK DIGITAL CURRENCY

The functions of a central bank in general are to maintain the central bank money reserves issued to the commercial banks and the fiat currency to be made available to the public via commercial bank branch counters and the automated teller machine. The commercial bank in turn maintains the

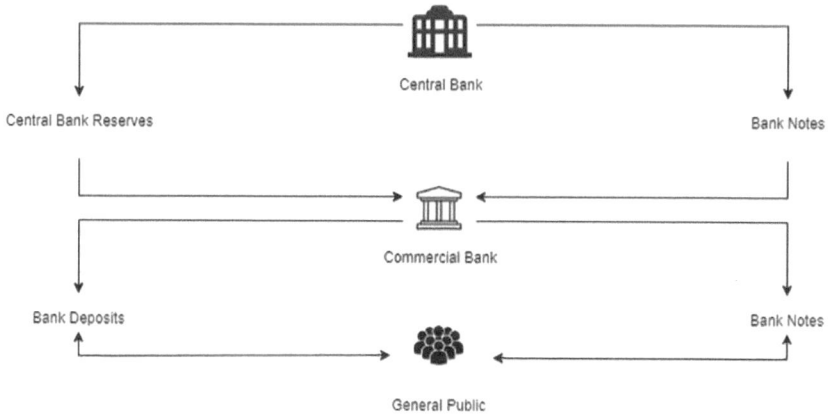

Figure 11.1 Financial structure in a country.

deposits of the public, and any settlement between the commercial banks happens with the central bank reserve maintained at the central bank (Deloitte, 2020) (Figure 11.1).

CBDC could be classified into two types:

- Wholesale CBDC
- Retail CBDC

11.2.1 Wholesale CBDC

A wholesale CBDC is a tokenized form of a central bank reserve which the commercial banks maintain with the central bank. This digital token could be used for bilateral settlement between the commercial banks, and the settlement of the token could be registered immutably in a decentralized private permissioned ledger, with each commercial bank maintaining a node in the network. This enables the transaction to be more transparent and eliminates the end-of-day reconciliation of the book of records between the commercial banks and the central bank. Also, to ensure privacy, the transactions would be visible to the commercial bank in which they were part of.

11.2.2 Retail CBDC

A retail CBDC is an e-money which the consumers would use to transact and which could be made centralized, wherein the individual consumers have an account with the central bank. However, this would result in the disintermediation of commercial banks, thus resulting in the destabilization of the existing financial structure. Hence, there are design choices which could be considered to include the commercial bank and other fintech providers to innovate and utilize the disbursement of the CBDC to the end consumers. The important aspect to be considered while designing the retail CBDC is anonymity as we are seeing it as a potential replacement to the physical cash.

11.3 CENTRAL BANK DIGITAL CURRENCY – DESIGN CHOICES

CBDC could be designed as follows:

- Account-based CBDC
- Token-based CBDC

11.3.1 Account-based CBDC

The account-based CBDC model ensures that the transaction initiated is linked to an owner. The transaction is approved based on the verification of user identities by the originator and the beneficiary. This resembles the system that we currently use; it makes CBDC non-anonymous but makes it more transparent and complies with the policies for anti-money laundering and other non-illicit activities. The non-anonymity nature of this design could demotivate wider acceptance of CBDC as consumers would prefer the CBDC to be a replacement to cash and hence remain anonymous (Deloitte, 2020) (Figure 11.2).

11.3.2 Token-based CBDC

The token-based CBDC model provides anonymity to the transactions, and the transaction is verified using public key cryptography and digital signature. The token-based design provides more accessibility to the end

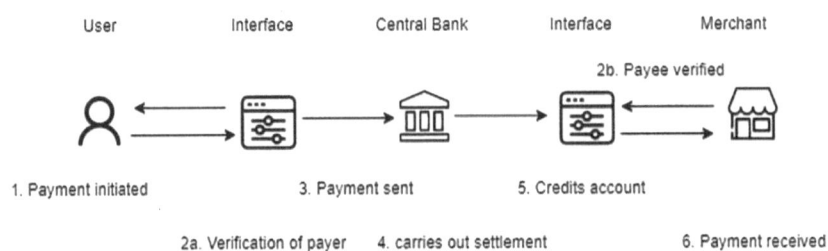

Figure 11.2 Flow in an account-based CBDC.

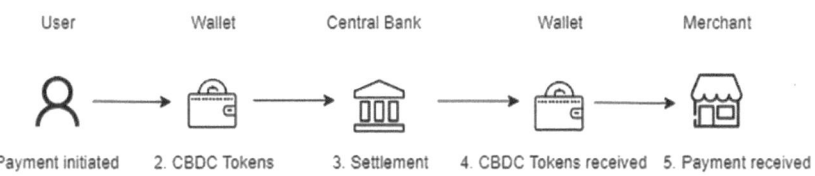

Figure 11.3 Flow in a token-based CBDC.

consumers as the CBDC remains bearer in nature and helps enable financial inclusion as it does not mandate the consumers to have an account. The design could also be made to work in the offline mode, thus enabling access in non-network areas (Deloitte, 2020) (Figure 11.3).

11.4 USE CASES INVOLVING WHOLESALE CBDC

11.4.1 DVP settlement of tokenized securities with CBDC

The tokenized form of the financial assets, either listed or unlisted securities, could be represented and made available in distributed ledger technology (DLT). The cash leg for this settlement must be a central bank money (CeBM), which would be tokenized to represent a W-CBDC and made available in the DLT. Upon initiation of the transaction, the tokenized securities would be settled upon the receipt of the W-CBDC in DLT in a delivery versus payment (DVP) manner (France, 2021) (Figure 11.4).

However, implementation of the W-CBDC via DLT could also leverage existing legacy systems and could be made interoperable with the DLT. The securities that are currently being settled in the legacy systems can

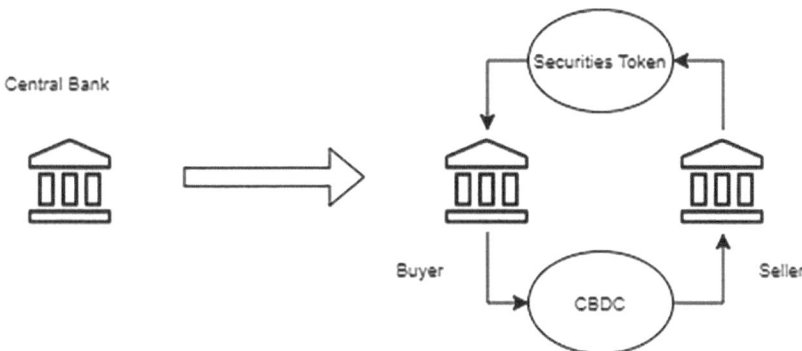

Figure 11.4 CBDC in the settlement of tokenized securities.

continue to get settled against a CBDC transferred on DLT (France, 2021) (Figure 11.5).

1. Delivery of securities in the legacy system can trigger the transfer of CBDC in DLT.
2. Transfer of CBDC in DLT can trigger the settlement of securities in the legacy system

11.4.2 Multi-CBDC for cross-border transactions

Cross-border transactions always do come up with high transaction costs due to the presence of multiple intermediaries and continue to be a woe for consumers transferring money from one country to another. Also, with respect to banks, they entail high transaction fees and operational inefficiencies.

This is where mCBDC (multi-CBDC) could be deployed by two different jurisdictions and agrees to transact on a common blockchain protocol (France, 2021) (Figure 11.6).

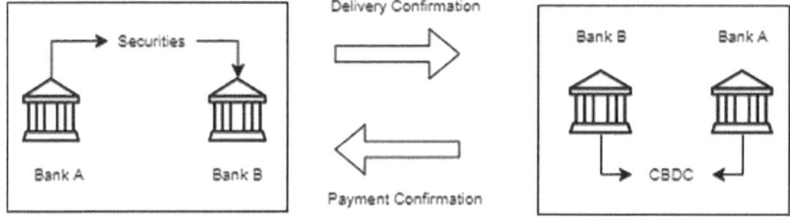

Figure 11.5 Integration of CBDC with the existing legacy applications.

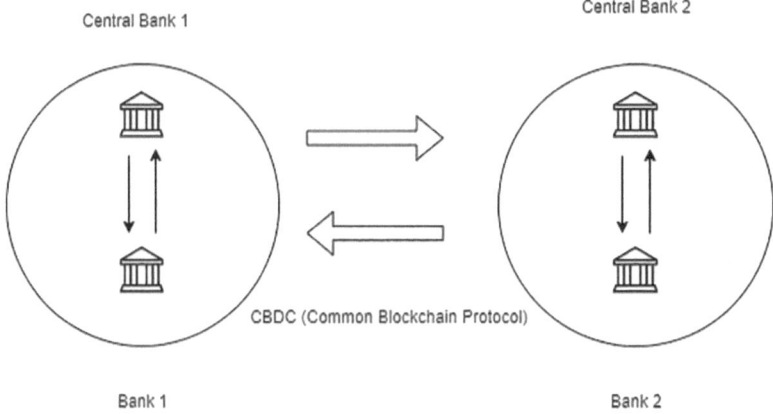

Figure 11.6 CBDC in cross-border transaction.

mCBDC arrangement could help the cross-border transaction in the following ways:

- mCBDC arrangement could enable a faster and secure settlement for the FX transactions.
- mCBDC increases transparency and reduces both processing time and costs.
- Implementation via DLT offers transaction traceability and visibility to all participants.
- mCBDC reduces the need for reconciliation.
- mCBDC enhances the security and flexibility of cross-border settlement.

11.5 IMPORTANCE OF ISSUING RETAIL CBDC

A decline in the usage of cash has led many central banks across the world to consider the usage of e-money for retail and domestic transactions. CBDC provides the public with access to CeBM. The reliance on the payment system dominated by private players could be mitigated and help foster a competitive environment. CBDC also helps in passing on the policy benefits directly to the consumers; any change in the repo rate takes time to reach the end consumers as it must go via the commercial banks. However, the CBDC could immediately reflect the policy rates announced by the central bank (Bank, 2021).

11.5.1 Design considerations for retail CBDC

The CBDC must entail the involvement of businesses, households, payment service providers and the central bank (Thailand, The Way Forward for Retail Central Bank Digital Currency in Thailand, 2021) (Figure 11.7).

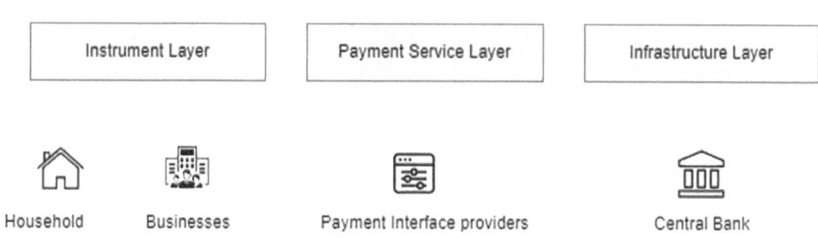

Figure 11.7 Retail CBDC – design.

Key functional attributes to be considered in the design of CBDC:

- Easy access
- Safety and security
- Interest bearing
- Convertibility
- Interoperability
- Settlement finality
- Minimization of financial risk

11.5.2 Direct CBDC

In this type, the CBDC would be directly issued to the end consumers by the central bank and no financial intermediaries would be involved. The CBDC issued would be a direct claim on the central bank (Settlements, 2020) (Figure 11.8).

11.5.3 Indirect CBDC

This is like the existing two-tier financial architecture that exists in the current landscape. The commercial bank would maintain the independent ledgers which are then connected to the wholesale ledger of the central bank. The end consumers would still have a claim on the commercial banks (Settlements, 2020) (Figure 11.9).

11.5.4 Hybrid/synthetic CBDC

The existing financial structure of the two-tier architecture is retained in the hybrid model. However, the claim on the CBDC would still be with the central bank and the commercial bank intermediaries would be responsible for the client onboarding and the KYC process (Settlements, 2020) (Figure 11.10).

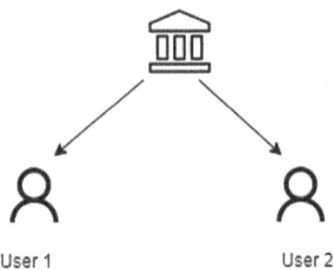

Figure 11.8 Direct CBDC – structure.

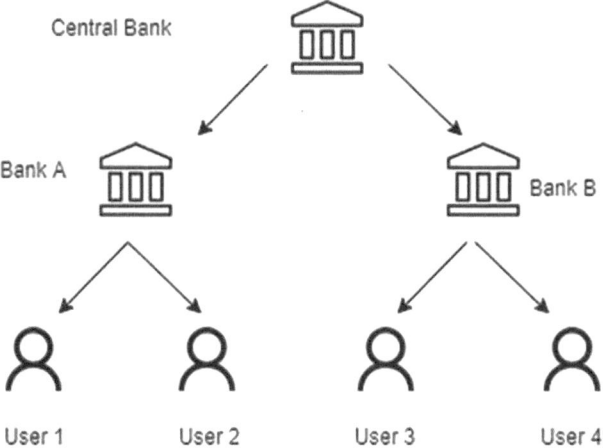

Figure 11.9 Indirect CBDC – structure.

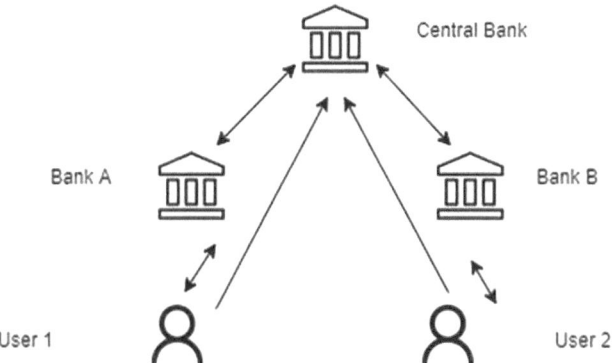

Figure 11.10 Hybrid CBDC – structure.

11.6 APPLICATION OF DLT FOR CBDC IMPLEMENTATION

Blockchain/DLT technology could bring varied advantages in the implementation of CBDC such as the following (Consensys, 2020) (Figure 11.11):

Programmability: The programmability feature of the blockchain could usher in innovation in the payment sector with the users having the control to bundle legal agreements into payments, thus enforcing the law via code.

Decentralized: The network is distributed across multiple nodes as the ledger is being shared across the nodes to maintain trust and transparency, thus reducing the operational and infrastructure-related costs.

Trust: With the participation of multiple nodes held by the central bank, commercial banks and regulatory bodies, the blockchain infrastructure embodies trust in the system, which is distributed and not concentrated.

11.6.1 Incorporation of blockchain for CBDC – advantages

Blockchain/DLT enables a peer-to-peer transfer of digital currency. When the currency is maintained as a token in the DLT along with other assets maintained as tokens, it enables a seamless transaction across assets and a DVP scenario.

The currency and the asset change its status during its lifetime, like a change in ownership and a change in the status based on certain events. These status changes could be effectively captured via smart contracts as they enable programmability (R3, 2020) (Figure 11.12).

11.6.2 Wholesale CBDC

The distributed network would compromise a node each from the commercial banks participating in the real-time gross settlement (RTGS) network and a node for the central bank. The transactions initiated by the commercial bank would remain privy to the only parties participating in the transaction and to the central bank which acts as a validator.

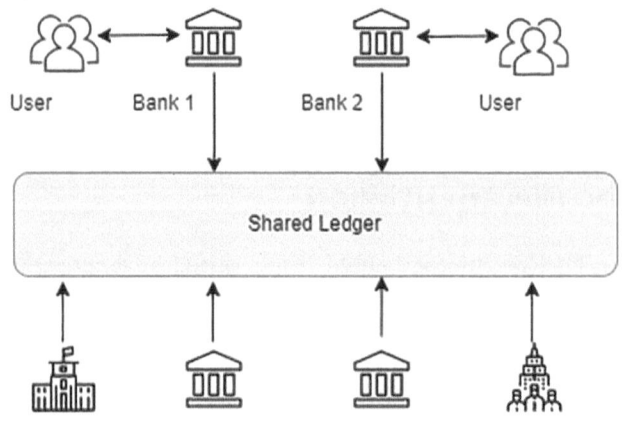

Figure 11.11 CBDC using DLT/blockchain.

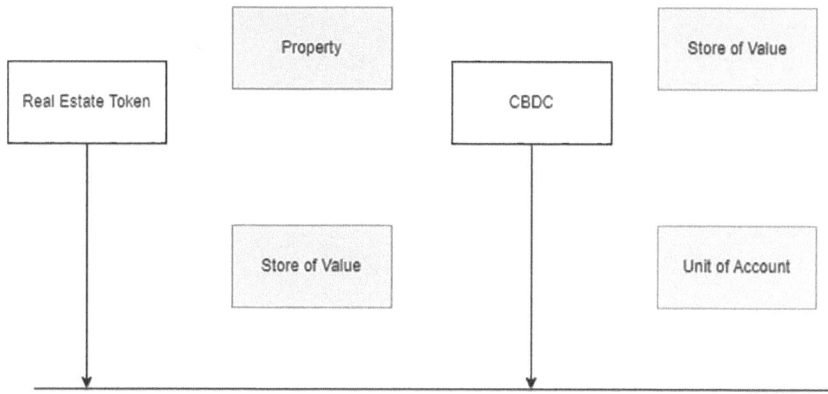

Tokenized Ecosystem (Transfer of value using CBDC against tokens representing Atos)

Figure 11.12 Payment of CBDC against a tokenized asset.

11.6.3 Retail CBDC

The distributed network could be implemented as a public network as it would be easy for the users to join the network via a wallet and utilize the CBDC. The group of trusted parties, in this case the bank, could be made as the validators. However, scalability and other technological improvements need to be considered before deploying this solution.

11.7 GLOBAL PROJECTS ON CBDC

11.7.1 Project Stella

Initiated by: European Central Bank and the Bank of Japan
  Use Case: Cross-border payment system
  Description: Conducted a cross-border payment transaction between the ledgers of the Bank of Japan and European Central bank by utilizing the interledger protocol mechanism and hashed time lock contracts (HTLC) and involving the connectors to achieve interoperability and to synchronize payments and lock funds along the payment chain (Bank, 2019).

11.7.2 Project Khokha

Initiated by: South African Reserve Bank (SARB)
  Use Case: Wholesale CBDC
  Description:

- Using tokenized South African rand on DLT for wholesale interbank settlement.

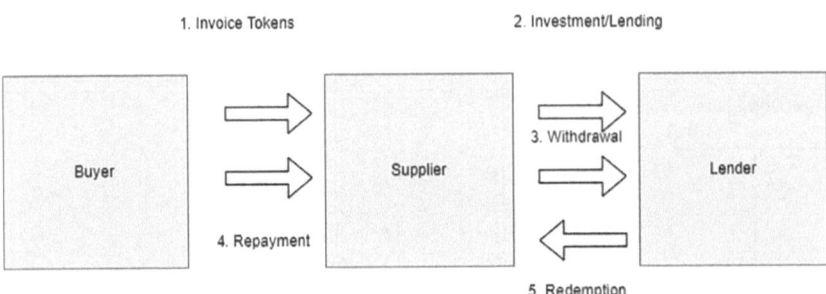

Figure 11.13 Project inthanon: financing via CBDC.

- Current RTGS settlement happens sequentially, and implementation of DLT could significantly reduce the time taken to settle payments across banks (Consensys, Project Khokha: Pushing the Limits of Interbank Payment Settlement with Blockchain, 2019).

11.7.3 Project Inthanon

Initiated by: Bank of Thailand
  Use Case: Invoice tokenization and financing workflow
  Description:
  The supplier posts the collateralized token of the invoice with the value in Thai Baht generated based on a smart contract to obtain financing from the lenders. The lender's CBDC wallet is connected to the application when the transaction is initiated to buy either full or partial tokens; then CBDC tokens from the lender's wallet are debited and credited with the invoice tokens.
  This way business can raise investment with the conversion of invoices into tokens (Thailand, 2018) (Figure 11.13).

- The supplier submits the invoice tokens represented in That Baht (CBDC) to the lender.
- The lender makes the investment.
- The supplier makes the withdrawal of the invoice tokens and obtains the repayment from the buyer.
- The lender redeems the invoice tokens with the supplier.

11.8 RISKS OF NOT IMPLEMENTING CBDC

- Financial intermediaries may choose private assets like cryptocurrencies/stable coins for tokenized settlement as they provide direct settlement on ledger and the CeBM is still not tokenized enough to various denominations.

- **Systemic Risk:** Non-implementation of CBDC could still maintain the systemic risk due to the concentration of liquidity and credit risks in the payment systems as banks and financial institutions remain as the single source for doing payments.
- Depositors would remain as vulnerable as their deposits may be lost in case of any liquidation of the banks. The banks do have deposit insurance in place, but history has time and again proved that depositors would be put under risk in case of any unforeseen circumstances (Kenya, 2022).

11.9 CONCLUSION

It is quite apparent that we are witnessing a radical change in the evolution of the new payment system and the digital currency powered by technological advancement, which is expected to remove the friction and hindrances in the existing payment infrastructure. The vision of a global and seamless currency, which is interoperable and seamless across multiple nations for payment, is a possibility in the future.

However, the immediate goals for the central banks of the nations are to identify the right use case, be it wholesale CBDC or retail CBDC, and identify the design architecture to implement and forge partnerships with the financial/non-financial firms and technology companies to jointly work together on the outcome desired.

There are still a lot of unknowns that need to be cleared in terms of the approach to alleviate the fears of disintermediation of banks, scalability of the solution and innovative use cases that could be built on top of CBDC. DLT with its advancement and the potential would be an appropriate fit in terms of helping to build a robust and secure framework for the roll-out of the CBDC.

REFERENCES

Bank, E. C. (2019). *Project Stella – Synchronized Cross-Border Payments.* Retrieved from https://www.ecb.europa.eu/paym/intro/publications/pdf/ecb.miptopical190604.en.pdf.

Bank, S. N. (2021). *Retail CBDC Purposes and Risk Transfers to the Central Bank.* Retrieved from https://www.snb.ch/n/mmr/reference/working_paper_2021_19/source/working_paper_2021_19.n.pdf.

Consensys. (2019). *Project Khokha: Pushing the Limits of Interbank Payment Settlement with Blockchain.* Retrieved from https://smallake.kr/wp-content/uploads/2020/10/Project-Khokha-Case-Study-Client-Ready-JUNE-2019.pdf.

Consensys. (2020). *Central Banks and the Future of Digital Money.* Retrieved from https://consensys.net/solutions/payments-and-money/cbdc/.

Deloitte. (2020). *Are Central Bank Digital Currencies (CBDC) the Money of Tomorrow?* Retrieved from https://www2.deloitte.com/content/dam/Deloitte/lu/Documents/financial-services/Banking/lu-are-central-bank-digital-currencies.pdf.

France, B. (2021). *Wholesale Central Bank Digital Currency Experiments with the Banque de France.* Retrieved from https://www.banque-france.fr/sites/default/files/media/2021/11/09/821338_rapport_mnbc-04.pdf.

Kenya, C. B. (2022). *Discussion Paper on Central Bank Digital Currency.* Retrieved from https://www.centralbank.go.ke/uploads/discussion_papers/CentralBankDigitalCurrency.pdf.

R3. (2020). *Central Bank Digital Currency: An Innovation in Payments.* Retrieved from https://www.r3.com/wp-content/uploads/2020/04/r3_CBDC_report.pdf.

Settlements, B. f. (2020). *The Technology of Retail Central Bank Digital Currency.* Retrieved from https://www.bis.org/publ/qtrpdf/r_qt2003j.pdf.

Thailand, B. o. (2018). *Project Inthanon.* Retrieved from https://www.bot.or.th/Thai/PaymentSystems/Documents/Inthanon_Phase1_Report.pdf.

Thailand, B. o. (2021). *The Way Forward for Retail Central Bank Digital Currency in Thailand.* Retrieved from https://www.bot.or.th/Thai/digitalcurrency/documents/bot_retailcbdcpaper.pdf.

Chapter 12

Blockchain technology in disaster management

A high-level overview and future research dimensions

Sreelakshmi S. and Vinod Chandra S. S.

University of Kerala

CONTENTS

12.1 INTRODUCTION

Natural hazards such as floods, storms, volcanic eruptions, earthquakes, and landslides have unpredictably affected the earth in a very tragic format. Proper management of disasters and pandemics is a difficult task. It consists of recording, processing, storing, and distributing information to the government and public to make effective decisions, such as allocating funds, initiating rescue operations, providing food and shelter, medical and rehabilitation centers, transportation facilities, water and power supplies, and other necessities. Many disaster relief agencies and plans are available but do not work correctly because of these challenges. There is a considerable scope for advanced technologies to overcome these difficulties and provide a secure way to store personal identification information and better security. One of the critical requirements for the future system is to store information and data safely and make the transaction secure and anonymous because securing the information against unauthorized access points is much more critical. Blockchain (BC) is the fastest-growing digital technology. It offers a robust self-regulating, self-monitoring, and cyber-resilient data transaction operation, assuring the facilitation and protection of a truly efficient data exchange system. BC is a shared-distributed database, and every node on a network can share, but no one controls it entirely. Encryption and

DOI: 10.1201/9781003282914-12

public are two fundamental features of the BC, and it uses private and public keys to provide better security. The objectives of BC-oriented disaster management are (i) to validate the information obtained from various sources, (ii) to integrate the information systems that work independently to achieve transparent, reliable, and consistent information, (iii) to protect and secure the identity and associated disaster-related information when issuing victim certificates, and (iv) to provide a cost-effective disaster management solution for operational and administrative purposes. As part of quick and efficient post-disaster recovery systems, some US states adopt BC-oriented technologies to better reconstruct buildings and infrastructure in cases of sudden climate change and unpredicted weather disasters. This proposed chapter critically evaluates the existing BC-oriented disaster management systems and their future research scope.

A disaster is a significant issue happening over a period that causes widespread human and socio-economic loss, which exceeds the capability of the victims or area to cope using its resources. Disasters can be caused by natural, artificial, and technological hazards and several aspects that influence the disclosure and vulnerability of a community. Natural disasters such as floods, storms, volcanic eruptions, earthquakes, and landslides have unpredictably affected the earth in a very tragic format. Proper management of disasters and pandemics is a difficult task. Disaster management aims to reduce the risk posed by actual and potential disasters. At present, disaster management has become more complex and specialized because of the increasing rate of extreme catastrophes and crises in almost every area. Reliable communication infrastructures are one of the major requirements of any disaster management plan (Ilbeigi et al., 2022). According to Hunt et al., the temporary running infrastructures gave centralized solutions only. The need for decentralized, secure technology remains highly important. All disaster management phases generate large amounts of data due to the service transactions and aid efforts that take place during disaster management. The use of BC technology in disaster management can reduce corruption, facilitate and accelerate the formation of partnerships between disaster relief agencies, deliver verified and timely disaster communications, improve the allocation of vital resources, and enable secure access to the valuable data that are produced during response and recovery operations (Hunt et al., 2022).

The major contributions of this chapter can be summarized as follows:

1. The possibility and benefits of using BC technology for enhancing the disaster management domain have been explored.
2. Various areas of application in disaster management where BC can be applied are presented.
3. Various challenges in the use of BC technologies in the disaster management domain have been explored and presented.

The remainder of this chapter is organized as follows:

12.2 BLOCKCHAIN TECHNOLOGY – AN OVERVIEW

BC technology is an emerging and rapidly growing technology, and it can be defined as a distributed database which contains digital transaction records or executed and shared public ledgers. On the other hand, it is a chain of blocks containing information that is entirely open to anyone, but no one controls it completely. Once we enter the data into the BC, they are complicated to change or delete. BC has worked efficiently and is successfully applied to several security applications throughout the world. It can bring a storm of revolution in the digital world in which all the transactions, both past and present, can be verified (Sekhar et al., 2019). This distributed ledger technology introduced as a backend infrastructure for the famous cryptocurrency Bitcoin provides a secured, immutable, tamper-proof data store. This trustless platform offers a transparent network infrastructure where the notion of trust can be taken from the organizations operating in a centralized manner. The trust will now be entrusted to the network where the transactions can be recorded in an immutable and tamper-proof way. This enables a group of non-trusting parties to collaborate and do business without any third-party authorization. The most exciting and invaluable property of BC is that it offers a decentralized data store, where every peer in the network keeps a copy of replicated data. The network manages the replication, sharing, and synchronization of the data across the peers through some consensus mechanism (Jennath et al., 2020).

The BC consists of mainly two elements: user transactions and blocks. In these blocks, transactions are recorded in the correct sequence and it is made sure that they have not been manipulated. The main characteristic features of the BC technology are shown in Figure 12.1. The major features are

   a. **Ledger:** It can be seen as a database of transactions. BC builds a ledger to manage all registered transactions.
   b. **Immutability:** Blocks are used to create the immutable ledgers where the transactions are cryptographically secured using hashing and digital signatures.
   c. **Decentralization:** It guarantees scalability and high-level security.
   d. **Anonymity:** This ensures that the participants are hidden and no personal identifiable information is captured.
   e. **Improved Security:** As it is a decentralized ledger, there will be no single point of failure.

A smart contract is a digital code where the business logic is implemented and enforced in the BC network. It establishes a set of predetermined conditions agreed upon by the business stakeholders and will automatically execute the terms of the agreement as soon as predefined needs are met. Smart contracts are, therefore, enforced by computer protocols without any

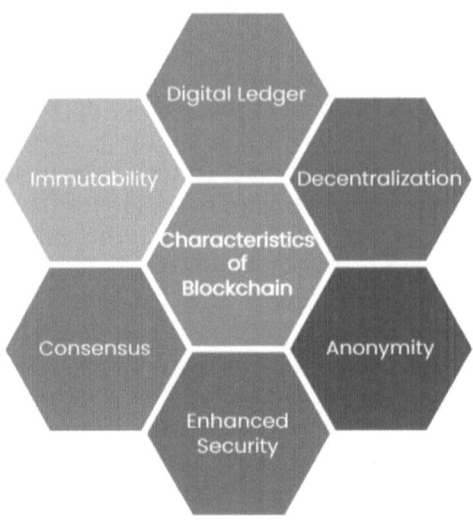

Figure 12.1 Characteristics of BC.

Table 12.1 Major myths about BC technology

Myth	Reality	Fact
BC and Bitcoin are the same.	Not exactly true. Bitcoin is a cryptocurrency application powered by BC.	BC technology can be used and configured for many other applications.
BC is better than the traditional database.	BC's advantages come with significant technical trade-offs that mean that traditional databases often still perform better.	BC is particularly valuable in low-trust environments where participants cannot trade directly or lack an intermediary.
BC is immutable or tamper-proof.	The BC data structure is append only, so data cannot be removed.	BC could be tampered with if >50% of the network-computing power is controlled and all previous transactions are rewritten, which is largely impractical.
BC is 100% secure.	BC uses immutable data structures, such as protected cryptography.	Overall BC system security depends on the adjacent applications, which have been attacked and breached.
BC is a "truth machine".	BC can verify all transactions and data entirely contained on and native to BC (e.g. Bitcoin).	BC cannot assess whether an external input is accurate or "truthful" – this applies to all off-chain assets and data digitally represented on BC.

human intervention and manual paperwork (L'Hermitte et al., 2021). The current BC systems may be classified roughly into public BC, private BC, and consortium BC (Zheng et al., 2017).

BC is a technology infrastructure introduced with the famous digital currency Bitcoin, which can record the transactions in a tamper-proof ledger and share the updated copy of the ledger with all the parties involved. Such a mechanism can eliminate the notion of centrality and take out the organizations' trust in a mathematically verifiable technological infrastructure. There are several myths and rumors around BC technology, for example, BC is another name for Bitcoin, BC is 100% secure, and BC is always better than traditional centralized databases, to name a few. Some of the common myths about BC and the reality and facts are given in Table 12.2.

Table 12.2 Expected performance and existing implementations for BC-enabled use cases

Use- Case	Type of BC	BC-powered solution
Contact tracing	Public Consortium	In the event of contact tracing, BC-powered decentralized models ensure better security and it permits users to hold on to the possession of information.
Disaster relief insurance	Consortium Private	In the insurance field, the BC-powered self-regulating decentralized financial systems can drive donors and produce better confidence.
Victim's information sharing	Consortium Private	BC-powered digitized healthcare domains allow the users to have secure data sharing in an efficient and effective manner.
Immigration and emigration process	Consortium Private	BC-enabled decentralized immigration ecosystems provide better collaboration among national and international level transmissions.
Supply chain management	Consortium Private	BC ensures the self-monitoring and trustworthy supply chain management system that strengthens the faster and more cost-efficient operations.
Automated surveillance and contactless delivery	Public Private	BC powers unmanned aerial vehicles (UAVs) and robots, and it is considered as the better strategy to self-regulate security surveillance and contactless delivery of essential things.
Online education and secure certification	Public Private	BC-enabled decentralized platforms for institutions can transform the way education is imparted online, and globally verifiable certificates are issued.
Manufacturing management	Public Private	BC-enabled technology has the potential to support the expansion of integrated trustworthy collaboration and better control monitoring systems for production and manufacturing with unprecedented precision.

(Continued)

Table 12.2 (Continued) Expected performance and existing implementations for BC-enabled use cases

Use- Case	Type of BC	BC-powered solution
e-Government	Consortium Private	BC-powered transparent distributed databases have the capability to provide better security and interaction power among citizens and government bodies.
Agriculture and food distribution	Consortium Private	BC nourishes the digital markets to be more trustworthy and enables realistic solutions to operating and distributing the food. Also, it certifies the transparency with anonymity to the stakeholders.

12.3 BLOCKCHAIN FOR DISASTER MANAGEMENT

It has found many demerits in implementing a centralized system for disaster management. BC-enabled systems provide a decentralized, safe, fast, potent, and sustainable disaster response mechanism (Poonia et al., 2021). Fruitfully contributing to the rescue guidelines is an effective solution for narrowing the devastating conclusions of the disaster situation. One of the crucial parts of the rescue guidelines is emergency management, which tactically relates the planning, technology, management, and science for handling the relief plan during the disasters (Drabek et al., 1991). The disaster management policy of any nation is mainly focused on good security as it is also a part of national security. A few days back, the website of the Indian disaster management authority was hacked by someone, and then the hackers sent some messages, posted something, and changed the display picture and name. The main challenge occurring in the case of a rescue plan is theft identification, which is spread quickly—stealing others' titles and other essential details for copious motives without their knowledge in this category. Most of the time, the people targeted by the identity thieves are unaware of their identity being stolen for months until they get a bill or a receipt of unauthorized transactions or a criminal case against their name (Panesir, 2018).

On-time refugee relief is a significant challenge for disaster management plans. Valid identification is essential for improving the quality of relief plans. Establishing a reliable connection between help requesters, information providers, and help givers is the critical enabler in such situations. One of the valuable sources of information during a disaster is crowdsourcing. Samir et al. proposed BC that enabled a decentralized trust management model used to facilitate cooperative anonymous help provisioning during disaster scenarios: community included members exchanging information about a disastrous situation, help requesters, and help providers. They assumed minimal infrastructure support where users could communicate to one of the local help centers. The help centers can be any emergency service building close to the disaster area, and the local help centers hold the chain files shareable to all connected community members (Alsalamah et al., 2020). The BC stores data on many computer nodes connected over the Internet, and its advancement will help the government and related authorities build proper disaster management (McIsaac et al., 2019).

Disaster victim identification (DVI) is a complex process where post-mortem identifying data, essentially fingerprints, DNA, and dental, are collected to compare antemortem data related to the missing person list. Even though there are many solutions and techniques to identify humans, they face several challenges. Thus, we need more sophisticated tools and methods for human identification during disasters in a timely fashion. Alsalamah et al. introduced a BC-enabled tool to facilitate building trustworthy, secure, and holistic ecosystems. It can disseminate siloed AM and PM data across systems, protecting against data breaches, redundancies, inconsistencies, and errors (Samir et al., 2019). Sobha et al. conducted a detailed study on BC in the disaster management domain. They concluded that these systems ensure users' privacy and that the information can be secured and kept in a tamper-proof manner (Sobha et al., 2019).

The Internet of Things (IoT)-enabled BC network proves its efficiency in uninterrupted data transmissions several times. Because it can overcome all the difficulties of a centralized transmission system, it allows for collecting the information provided by the devices in a decentralized manner while reducing the costs associated with maintaining large data centers (Aranda et al., 2019). After-disaster actions sometimes are even more important than actions performed during a disaster (Anjomshoae et al., 2017). One of the most critical after-disaster activities is the supply of medicines and essential nourishment products such as milk or even water that should be guaranteed for the disaster survivors. BC and IoT can help determine which closest warehouses can be accessed to maintain flows of goods and for how long to optimize new orders to providers and reroute in case of need (Tang, 2006; Shaluf, 2008). In addition, food distribution can benefit from these technologies by ensuring that food reaches the most needed victims through smartphone camera eyeball identification avoiding bad actors taking advantage of the chaos and getting more than their fair share of pretending to be victims (Adinolfi et al., 2005). IoT is a promising technology used in several applications, including disaster management. In disaster management, the role of IoT is so essential and ubiquitous and could be life-saving (Sharma et al., 2021). IoT-enabled BC technologies are already proving their efficiency in various domains. Ratta et al. (2021) conduct a detailed study on this hybrid technology in the healthcare domain. Sarbajna et al. (2021) find an IoT-enabled BC solution to the problem of immediate creation of traversal maps during disaster situations.

Supply chain management in disaster relief is a critical system responsible for designing, deploying, and managing the required processes to deal with current and future disaster events and managing the coordination and interaction of the operations with other competitive or complementary supply chains (Pour, 2021). The main cornerstones of a disaster supply chain are collaboration, communication, and contingency planning. Different parties pursue several distinct and, in some cases, conflicting strategic goals in contributing to disaster response (Vivek et al., 2009). The core challenge of a disaster supply network is a highly required decentralized database service like BC. A distinct number of supply chain models used BC technology as their core. Aranda et al. (2019) developed a BC-based management system named "vaccine blockchain", which can trace and manage the information in vaccine supply chains and provide valuable vaccine recommendation information to different users. Tian, 2016 integrated radio-frequency identification (RFID) and BC technology to develop a traceability system for the agricultural food supply chain. Recent studies focused on finding individual BC adoption behavior in India and USA and applied BC technology to the field of supply chains (Queiroz et al., 2019). In post-disaster recovery, the rebuilding process is a crucial and lengthy task. The prime concern of the productive disaster management system is to preserve a top level of perfection in the communication of vital information related to the area where it is to be distributed to government and relief agencies. BC technologies have made significant progress in recent years and are widely indicated to disrupt existing socio-technological ranges and workflows (Nawari et al., 2019).

The intrinsic features of distributed ledger make it a suitable model for implementation in disaster management systems, as evidenced by recent efforts. Connecting services tasked with delivering food, water, and other assistance often lacks transparency between different operations, complicated coordination between various parties, complex logistic planning, delivery delays, shortages, and waste of resources (Guillot, 2018). Rohr (2016) states that disaster situations necessitate absolute transparency, which only distributed networks can provide (Zambrano et al., 2018). BC can thus act as the central system of all operations by connecting all involved parties and enhancing the convenience of communication while upholding a secure protocol over the network. Panesir proposed a model that suggests an integrative model over a BC network that connects government bodies, medical suppliers, shelter providers, relief aid, telecom service providers, residents, and transportation providers (Panesir, 2018). The transparency and automated dissemination of recorded information characteristics of BC can negate the requirement of human involvement and arduous paperwork for sanctioning approvals. They can further streamline other managerial operations (Guillot, 2018). 'Building Blocks' is a BC initiative launched in January 2017 by the World Food Program (WFP). This initiative targets the

Azrap refugee camp in Jordan to increase the overall efficiency of the cash-based transfer scheme in several processes, such as refugee registration and financial accounting. Initial development and deployment were achieved in under 5 months. Building Blocks were meant to augment and increase the efficiency of the existing digital infrastructure of WFP's processes for beneficiary information registration and payment mechanisms by streamlining backend processing. Chakrabarti et al. (2019) proposed a BC-based incentive fully decentralized scheme for a DTN-enabled smartphone leveraged post-disaster communication network that uses the Bitcoin currency system to incentivize nodes for cooperation. It involves an integrative synthesis of BCs, digital databases, and biometrics and is estimated to have saved a WFP monthly cost of $150,000 by eliminating 98% of bank-related fees (Panesir, 2018). In contrast, the cash system used by the beneficiaries was unchanged.

The success of a disaster response network is based on collaboration, coordination, sovereignty, and equality in relief distribution. That is, the need for a trust-based communication system is required for accurate disaster management plans. Farinaz and Gheorghe (2021) proposed a BC-enabled transparent, secure, and real-time information exchange system as a disaster-aid network and offered various future research paths. Kalla et al. conducted a detailed study that provides a high-level discussion of how BC can support numerous use cases, including disaster management plans. Table 12.3 intends to reflect the performance of expected BC-powered systems and the implementation of existing methods for use cases (Kalla et al., 2020).

This literature review indicates that like any other exponential and disruptive technology, BC also found several use cases in disaster management. While disaster management is a highly human-involved procedure, technological innovations can assist various stakeholders involved in making informed decisions that will reduce the impact of such mishaps. The BC-enabled information systems can securely record sensitive information and help post-disaster management tasks such as contract tracing and the disbursement of disaster relief funds/insurance. The elimination of the third parties from the system will improve the efficiency of such activities, and the immutable ledger may act as a proof of transactions for different stakeholders involved. Some of the significant use cases, along with the type of BC and the corresponding solutions for using BC for disaster management, are outlined in Table 12.2.

12.4 MAJOR FINDINGS AND SUGGESTIONS

BC technologies can assist in increasing the speed and level of availability of needed materials and services that may be required during or in the aftermath of disasters for rescue and recovery (Badarudin et al., 2020). BC technologies are not limited to virtual currency transactions; they can also be applied to several information management systems like the health domain, disaster management sector, supply chain, etc. The integrated model is better than a single model. BC and IoT technologies have the potential to improve the speed, security, transparency, or efficiency to provide resilience for all infrastructures necessary to provide relief which may have their local computer systems flooded or unusable due to the disaster. According to the United Nations Office for Disaster Risk Reduction (UNDRR), the disaster management plan involves five major phases named prediction, early warning, emergency relief, rehabilitation, and reconstruction (Figure 12.2).

During the prediction phase, mitigation and preparedness activities are established, and it is essential to create BC-enabled decentralized humanitarian autonomous organizations to improve cooperation on collective trust. It can also help automate regulatory compliance requirements on safety and eliminate duplication efforts in identification processes while permitting encrypted updates of evacuates in real time. In the early warning phase, BC with IoT can help to ensure, on one hand, that the information comes from identified institutions and, on the other hand, that data provided are sent by identified IoT devices. Life preservation and subsistence of victims are the main focus of the emergency relief phase, which may take more time. Smart contracts are scripts stored in the BC that improve the operational efficiency. They provide security and integrity for IoT data. They are crucial in this stage to achieve fast and accurate relief. The WFP has already started to trial BC technology as a payment system and alternative tool for distributing cash-based transfers. In the rehabilitation phase, BC-enabled systems will secure victims' identities and clinical data and reduce fraud in activities. Finally, in the reconstruction phase, automated validation of claims can be performed through BC, increasing the security and efficiency of the process. This way, real-time automatic claim processing, eligibility verification,

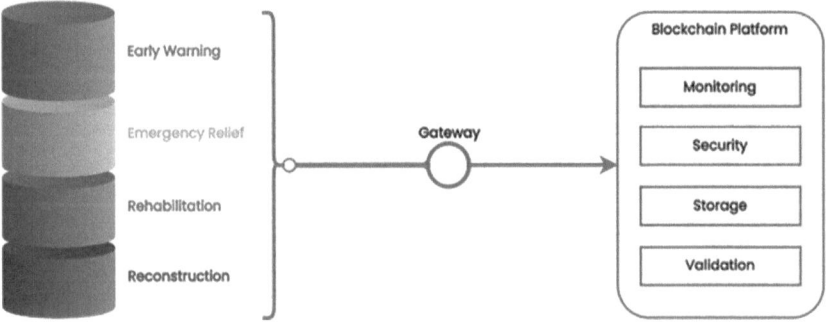

Figure 12.2 A conceptual framework for BC-based disaster management.

and preauthorization could become a reality. In addition, researchers could access different subsets of the stored data for research with relevant consent. In this stage, issues such as piracy prevention, completion of the chain of custody, and establishing a perimeter of trust can benefit dramatically from BC and IoT implementation.

12.5 OPEN CHALLENGES AND FUTURE OUTLOOK

BC, the technology behind the famous cryptocurrency Bitcoin, transforms how businesses function. Even though there are some early adopters of BC in various sectors such as healthcare, land registry, and supply chain, there needs to be a significant level of development to make BC a widespread adoption. The scalability and interoperability of BC technology are substantial concerns for enterprises and other stakeholders while implementing BC technology for their businesses. This is true for using this disruptive technology for disaster management activities. Some of the significant challenges in using BC in the disaster management domain, but not limited to, are outlined below:

Interoperability: Ensuring the proper interoperability among disaster management teams can be a challenge. On the other hand, exchanging the huge volume of information to the varied providers is very difficult due to its large open nature.

Security Issues: Decentralization ensures more security, but some disadvantages are still associated with it. The data are distributed in a public ledger, which can cause privacy leakage.

Lack of Standardization: Proper standardization of protocols and technologies is very important in the case of disaster management. Aspects like what data, size, and format can be sent to the BC, and what data can be stored in the BC should be clearly defined.

Data Ownership and Accountability: Who will hold the data, who will grant permission to share people's private data, and who will have the ownership are the main questions.

These are not the only challenges associated with implementing BC technology in disaster management. The technology may be an enabler, but many processes need to be re-engineered to implement the technology effectively. There are numerous benefits such an implementation can bring, such as anytime, anywhere access of data, onboarding of all stakeholders who are involved in the disaster management activities in one network, recording of donations received for the disaster recovery activities, enhanced coordination of the volunteering efforts, better tracking of the distribution of relief services, and improved track and trace capabilities, to name a few.

BC technology is still in its infancy stages, and we need more developments and innovations to take full advantage of this most disruptive technology ever since the Internet. Identity management is one of the most crucial components of a BC-powered disaster management framework. There are BCs for managing the digital identities of any individual, such as Hyperledger Indy; more privacy-preserving decentralized identity management techniques need to be developed. This is one of the future challenges that should get significant attention from the researchers. For effective disaster recovery and relief operations, identity tracking capabilities also need to be developed and integrated with the identity management module. While smart contracts can automate the fund collection and release related to the post-disaster stages, the regulation and auditability of smart contracts remain a hindrance. There needs to be regulation and laws from the government on vetting these intelligent contracts before they can be deployed in the network as a minor bug in the smart contract may make the funds transferred to the illegitimate hands. Regulatory standards need to be developed for handling these types of illegal activities. Software flaws in BC need to be taken seriously as there were reports that there was a loss of $24 million last year due to this BC security issue. Therefore, developing security and auditing standards for smart contract development is another area worth exploring. As mentioned earlier, integrating BC technology with other exponential technologies such as artificial intelligence and IoT is essential for interoperable use cases to be implemented. There need to be significant research activities centered around improving the interoperability aspect of BC technology.

12.6 CONCLUSIONS

BC technology promises to improve humanitarian operations' efficiency, transparency, and accountability. Like any other disruptive technologies such as artificial intelligence, machine learning, and IoT, BC also finds several exceptional use cases in disaster management. It is evident that BC technology is still in its infancy stage, and extensive research and enhanced regulatory interference from the government are needed to take the technology to its full potential. This chapter discussed the underpinnings of BC technology and reviewed some relevant and recent works reported in disaster management that use BC technology for pre- and post-disaster management activities. This chapter also outlined a conceptual framework for BC-based disaster management that uses the immutable and tamper-proof ledger for early warning, emergency relief, rehabilitation, and reconstruction activities. Even though a few approaches are reported for the same, there need to be significant research and developments in these areas to use the technology to its full potential. This chapter also presents some exciting research dimensions and open research problems for potential interdisciplinary researchers to investigate.

REFERENCES

Adinolfi, C., Bassiouni, D. S., Lauritzsen, H., & Williams, H. R. (2005). *Humanitarian Response Review*. Available online: https://digitallibrary. un.org/record/556468?ln=en

Alsalamah, S., & Nuzzolese, E. (2020). Promising blockchain technology applications and use case designs for the identification of multinational victims of mass disasters. *Frontiers in Blockchain*, 34.

Anjomshoae, A., Hassan, A., Kunz, N., Wong, K.Y. and de Leeuw, S. (2017). Toward a dynamic balanced scorecard model for humanitarian relief organizations' performance management. *Journal of Humanitarian Logistics and Supply Chain Management*, 7(2), 194–218. https://doi.org/10.1108/JHLSCM-01-2017-0001

Aranda, D. A., Fernández, L. M. M., & Stantchev, V. (2019). Integration of internet of things (IoT) and blockchain to increase humanitarian aid supply chains performance. In *2019 5th International Conference on Transportation Information and Safety (ICTIS)* (pp. 140–145). IEEE.

Badarudin, P. H. A. P., Wan, A. T., & Phon-Amnuaisuk, S. (2020). A blockchain-based assistance digital model for first responders and emergency volunteers in disaster response and recovery. In *2020 8th International Conference on Information and Communication Technology (ICoICT)* (pp. 1–5). IEEE.

Chakrabarti, C., & Basu, S. (2019). A blockchain based incentive scheme for post disaster opportunistic communication over DTN. In *Proceedings of the 20th International Conference on Distributed Computing and Networking* (pp. 385–388).

Drabek, T. E., & Hoetmer, G. J. (1991). *Emergency Management: Principles and Practice for Local Government*. Washington, DC: International City Management Association.

Farinaz, S. A. B. Z., & Gheorghe, P. N. M. (2021). A blockchain-enabled model to enhance disaster aids network resilience. *Romanian Cyber Security Journal*, 3(2), 77–87.

Guillot, C. *How Blockchain Could Speed Hurricane Disaster Relief, ThirtyK*. 2018. Available online: https://thirtyk.com/2018/07/31/hurricane-relief-blockchain/ (accessed on 20 April 2019).

Hunt, K., & Zhuang, J. (2022). Blockchain for disaster management. In *Big Data and Blockchain for Service Operations Management* (pp. 253–269). Cham: Springer.

Ilbeigi, M., Morteza, A., & Ehsani, R. (2022). An infrastructure-less emergency communication system: A blockchain-based framework. *Journal of Computing in Civil Engineering*, 36(2), 04021041.

Jennath, H. S., Anoop, V. S., & Asharaf, S. (2020). Blockchain for healthcare: Securing patient data and enabling trusted artificial intelligence. *International Journal of Interactive Multimedia & Artificial Intelligence*, 6(3), 15–23.

Kalla, A., Hewa, T., Mishra, R. A., Ylianttila, M., & Liyanage, M. (2020). The role of blockchain to fight against COVID-19. *IEEE Engineering Management Review*, 48(3), 85–96.

L'Hermitte, C., & Nair, N. K. C. (2021). A blockchain-enabled framework for sharing logistics resources during emergency operations. *Disasters*, 45(3), 527–554.

Liang, X., Shetty, S., Tosh, D., Kamhoua, C., Kwiat, K., & Njilla, L. (2017). Provchain: A blockchain-based data provenance architecture in cloud environment with enhanced privacy and availability. In *2017 17th IEEE/ACM International Symposium on Cluster, Cloud and Grid Computing (CCGRID)* (pp. 468–477). IEEE.

McIsaac, J., Brulle, J., Burg, J., Tarnacki, G., Sullivan, C., & Wassel, R. (2019). Blockchain technology for disaster and refugee relief operations. *Prehospital and Disaster Medicine*, 34(s1), s106.

Nawari, N. O., & Ravindran, S. (2019). Blockchain and building information modeling (BIM): Review and applications in post-disaster recovery. *Buildings*, 9(6), 149.

Panesir, M. S. (2018). *Blockchain Applications for Disaster Management and National Security* (Doctoral dissertation, State University of New York at Buffalo).

Poonia, V., Goyal, M. K., Gupta, B. B., Gupta, A. K., Jha, S., & Das, J. (2021). Drought occurrence in different river basins of India and blockchain technology based framework for disaster management. *Journal of Cleaner Production*, 312, 127737.

Pour, F. S. A. (2021). *Application of a Blockchain Enabled Model in Disaster Aids Supply Network Resilience* (Doctoral dissertation, Old Dominion University).

Queiroz, M. M., & Wamba, S. F. (2019). Blockchain adoption challenges in supply chain: An empirical investigation of the main drivers in India and the USA. *International Journal of Information Management*, 46, 70–82.

Queiroz, M. M., Telles, R., & Bonilla, S. H. (2019). Blockchain and supply chain management integration: A systematic review of the literature. *Supply Chain Management: An International Journal*, 25(2), 241–254.

Ratta, P., Kaur, A., Sharma, S., Shabaz, M., & Dhiman, G. (2021). Application of blockchain and internet of things in healthcare and medical sector: Applications, challenges, and future perspectives. *Journal of Food Quality*, 2021, https://doi.org/10.1155/2021/7608296.

Rohr, J. (2016). *Blockchain for Disaster Relief: Creating Trust Where It Matters Most*. Available online: https://www.digitalistmag.com/improving-lives/2017/11/23/blockchain-for-disaster-relief-creatingtrust-where-it-matters-most-05527536 (accessed on 24 June 2019).

Samir, E., Azab, M., & Jung, Y. (2019). Blockchain guided trustworthy interactions for distributed disaster management. In *2019 IEEE 10th Annual Information Technology, Electronics and Mobile Communication Conference (IEMCON)* (pp. 0241–0245). IEEE.

Sarbajna, R., Eick, C. F., & Laszka, A. (2021). DEIMOSBC: A blockchain-based system for crowdsensing after natural disasters. In *2021 3rd Conference on Blockchain Research & Applications for Innovative Networks and Services (BRAINS)* (pp. 17–20). IEEE.

Sekhar, S. M., Siddesh, G. M., Kalra, S., & Anand, S. (2019). A study of use cases for smart contracts using Blockchain technology. *International Journal of Information Systems and Social Change (IJISSC)*, 10(2), 15–34.

Shaluf, I. M. (2008). Technological disaster stages and management. *Disaster Prevention and Management: An International Journal*, 17(1), 114–126.

Sharma, K., Anand, D., Sabharwal, M., Tiwari, P. K., Cheikhrouhou, O., & Frikha, T. (2021). A disaster management framework using internet of things-based interconnected devices. *Mathematical Problems in Engineering*, 2021.

Sobha, G. V., & Sridevi, P. (2019). *Use Case of Blockchain in Disaster Management – A Conceptual View*. Greeley, CO: Aims Community College.

Tang, C. S. (2006). Robust strategies for mitigating supply chain disruptions. *International Journal of Logistics: Research and Applications*, 9(1), 33–45.

Tian, F. (2016). An agri-food supply chain traceability system for China based on RFID & blockchain technology. In *2016 13th International Conference on Service Systems and Service Management (ICSSSM)* (pp. 1–6). IEEE.

Vivek, S. D., Richey Jr, R. G., & Dalela, V. (2009). A longitudinal examination of partnership governance in offshoring: A moving target. *Journal of World Business*, 44(1), 16–30.

Zambrano, R., Young, A., Verhulst, S. (2018). *Blockchange Case Study: Connecting Refugees to Aid through Blockchain-Enabled ID Management: World Food Programme's Building Blocks*. 2018. Available online: https://blockchan.ge/blockchange-resource-provision.pdf (accessed on 14 June 2019).

Zheng, Z., Xie, S., Dai, H., Chen, X., & Wang, H. (2017). An overview of blockchain technology: Architecture, consensus, and future trends. In *2017 IEEE International Congress on Big Data (BigData Congress)* (pp. 557–564). IEEE.

Chapter 13

Interoperable blockchain systems

Radha Sridharan and Manju Sadasivan
CMR University

CONTENTS

13.1 INTRODUCTION

We live in a world that is increasingly interconnected. New technologies connect people and devices even more safely and efficiently. One of these technologies is the blockchain. The term 'blockchain', which is a chain of blocks, was suggested by Satoshi Nakamoto in 2008 [1].

DOI: 10.1201/9781003282914-13

Blockchain is a decentralized, distributed digital ledger of transactions that records information that cannot be changed and is evolving to become an integral part of our lives. We all know crypto assets as the first successful application of blockchain technology. However, the possible blockchain best use cases seem to be limitless. Digital identity, access control, ticketing, intellectual property, financing solutions, and the number of applications utilizing blockchain technology are growing at a rapid pace.

The technology of blockchain has evolved into four stages as shown in Figure 13.1.

Blockchain 1.0 has an electronic payment system or cryptocurrency application originated from the concept of distributed ledger technology. It eliminates the need of any central authority and implements consensus models to handle trust-related issues. It started with the Bitcoin network and had limited functionality catering only to the financial sector [25].

Blockchain 2.0, the second generation Ethereum technology, includes Bitcoin 2.0 protocols and provides smart contracts and Ethereum smart contracts. It enables developers to build distributed decentralized applications also called DApps and also implement self-executing contracts through smart contract technology. With these new features, the role of blockchain could secure and record the transactions intact as a permanent ledger. These blockchains remain isolated. The problem is that blockchains cannot interact with each other by nature. This means that the blockchain has no knowledge of what happens in other blockchains. However, the question is will different blockchains ever be able to interact with one another or will they remain working as silou solutions? The solution to all of these is blockchain interoperability. Interoperability is a key part of this ecosystem, and mass adoption would not happen without it. It is a great catalyst just like how the Internet was not mass adopted until transmission control protocol/Internet protocol (TCP/IP) came along.

With the advent of Blockchain 3.0, the importance of improved scalability, privacy, interoperability, and sustainability finally came to the forefront.

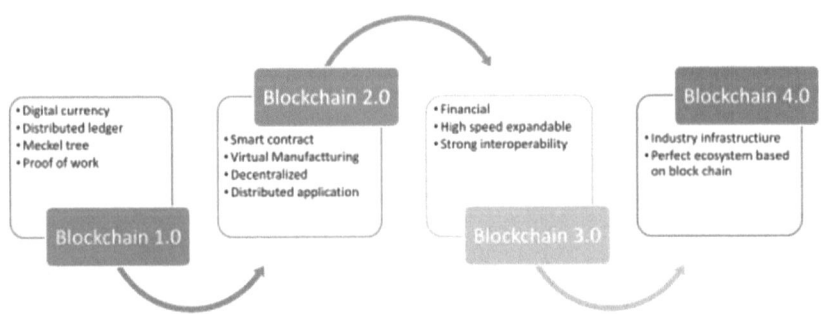

Figure 13.1 - Evolution of blockchain (Huang, Lin-Yun et al. 2005)

Blockchain 3.0 enabled distributed applications (dapps) to be implemented in several domains beyond financial applications or asset transfers. In the third generation of Blockchain 3.0, many projects were focusing on making sure that there is an endless smooth interconnection between blockchains and the convergence toward the decentralized applications was introduced [21].

Blockchain 4.0 includes industry 4.0 applications that focus on distributed ledger technology and real-life blockchain applications. Blockchain networks will not be mass adopted until interoperability comes into play. However, what is interoperability and what does it actually do?

13.2 WHAT IS BLOCKCHAIN INTEROPERABILITY?

One of the crucial features of blockchain technology that are emerging is interoperability. Interoperability is the ability for different hardware and software to exchange information freely without any restrictions. It is the communication technique between various blockchain systems. If there was no interoperability, then everyone would have to use the exact system to communicate. Therefore, in crypto, interoperability is the ability of different blockchains to exchange information freely.

Interoperability techniques enable people to gain access to information across different blockchain networks. Each blockchain stores different kinds of data and transactions. Implementation of interoperability will increase the communication rate of different blockchains. For instance, information can be transferred from an Ethereum blockchain to an EOS blockchain. Interoperability implementation will increase the communication rate of blockchains. Blockchain technology works on sharing and integration of data, which is more strengthened by making use of interoperability. This feature enhances blockchain transactions. By providing interoperability, the success rate of existing projects is increased to a great extent and also allows the advent of more emerging projects, mainly for fields where the value chain is important, like finance, aviation, healthcare, etc. [2].

Thus, the main aim of interoperability is to create a true decentralized network which is not limited to one blockchain but multiple blockchain networks [3].

Currently, different blockchain networks exist providing interoperability solutions to various other blockchain networks. Examples of projects that provide such solutions are Harmony, Polkadot, and Cosmos [4].

13.3 NEED FOR INTEROPERABILITY

Interoperability is really important. The success of blockchain technology depends on it, and many use cases would not be possible without it. Mass adoption definitely would not be possible without it. If there was no interoperability, then everyone would have to use the same blockchain to communicate, which makes very little sense. These days, everyone is creating his/her own blockchains which are built for different purposes and use cases. It is highly unlikely that there will only be one blockchain that everyone uses for everything, and it is probably incredibly inefficient to do so. Today, just like the pre-TCP/IP days of the Internet, these blockchain networks do not communicate with each other. They are all mostly siloed in their own ecosystems. Because everyone has different ideas and builds his/her projects differently for different use cases. Each crypto potentially has different consensus mechanisms, different semantics, and different hashing algorithms built with different languages. Some are private, and some are public. Therefore, it is no surprise that there is a huge range of different networks. This makes interoperability a complex problem to solve. However, if it is solved and invested in such projects, it will give great returns. Another reason is the fact that most people tend to think that they can pick the right project out of thousands and then fall into a self-confirming bias of believing that their project will go to the moon.

Blockchain interoperability generally has the ability to handle sharing states and transacting across different chains. Blockchains can be visualized as isolated databases, without proper interfaces for input or output of data. Blockchain interoperability could enrich and enhance use cases for blockchains like portable assets, payment versus delivery, and cross-chain oracle. To be precise, different blockchains would be abstracted, such that a user can readily manipulate all the functions without an accurate understanding of each blockchain.

The following are the prerequisites to achieve blockchain interoperability [27].

a. The presence of a cross-chain communication protocol that has the ability to transmit arbitrary data in a decentralized, trustless fashion.
b. Bridging set of blockchains through such a protocol.
c. Applications that can take advantage of a multiple blockchain approach, that is, Internet of blockchains to powered blockchain of blockchain applications.

13.4 WHAT IS THE CURRENT LANDSCAPE OF INTEROPERABILITY? – TECHNOLOGIES FOR BLOCKCHAIN INTEROPERABILITY

The first step toward blockchain interoperability has been to identify approaches that allow the exchange of cryptocurrencies. Such approaches are called cryptocurrency-directed interoperability approaches or public connectors. They comprise relays and side chains, notary schemes, and hashed time lock contracts. This approach showed the way for blockchain interoperability. Earlier, this approach was restricted to exchange of tokens between homogeneous blockchains. However, now there exists innovative blockchain interoperability methods, which include blockchain of blockchains and hybrid connectors. Hybrid connectors comprise blockchain-agnostic protocols, trusted relays, and blockchain migrators [5].

Among public blockchains, the most suited solutions for interoperability are sidechains combined with blockchain-agnostic protocols. They provide built-in interoperability between blockchains of the same platform. There was a great deal of difficulty in creating and maintaining a decentralized application using several blockchains. To resolve this, blockchain of blockchains was introduced. By incorporating variations to the solutions mentioned above, it can link blockchain of blockchains to other blockchains. Let us look at these technologies in detail.

13.4.1 Sidechains

Sidechains are an elegant way to overcome current limitations regarding scalability, interoperability, and governance in the blockchain ecosystem. Sidechain is a blockchain that is connected to the main chain. They can both interpret with each other through what is called a cross-chain communication protocol. The sidechain maintains its own ledger, and it can have its own consensus mechanism and native asset. Therefore, it has its own environment separated from the main chain. The three major components of sidechain are mainchain, sidechain, and cross-chain communication protocol. A sidechain allows interoperability between two existing blockchains [19].

The mainchain is one blockchain, and the other is called the sidechain. The mainchain maintains a ledger of assets. The connection between the mainchain and sidechain is provided through a cross-chain communication protocol. A two-way communication is provided by sidechains which transfer assets between the mainchain and sidechain [28]. Figure 13.2 shows the sidechain connected with the mainchain.

The construction of the sidechain model consists of two parts:

1. The sidechain consensus protocol—SCP
2. The two-way peg cross-chain transfer protocol—CTP

As shown in Figure 13.3, the sidechain consensus protocol governs how the network agrees on new blocks and therefore concentrates on the history of transactions [6].

The cross-chain transfer protocol determines how assets can be transferred from the mainchain to a sidechain and vice versa. Sidechains require

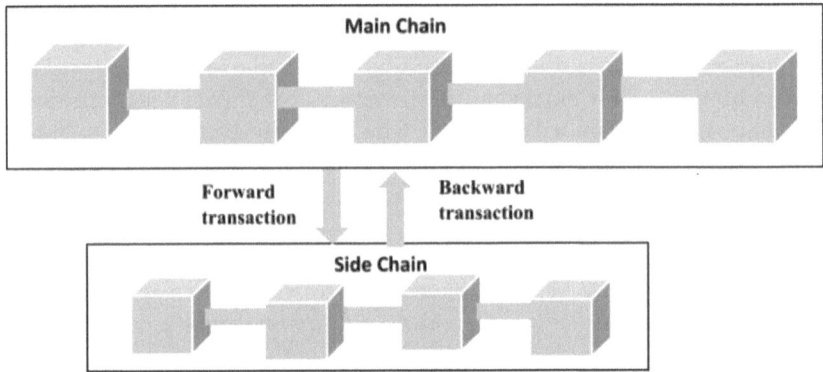

Figure 13.2 An example of a side chain

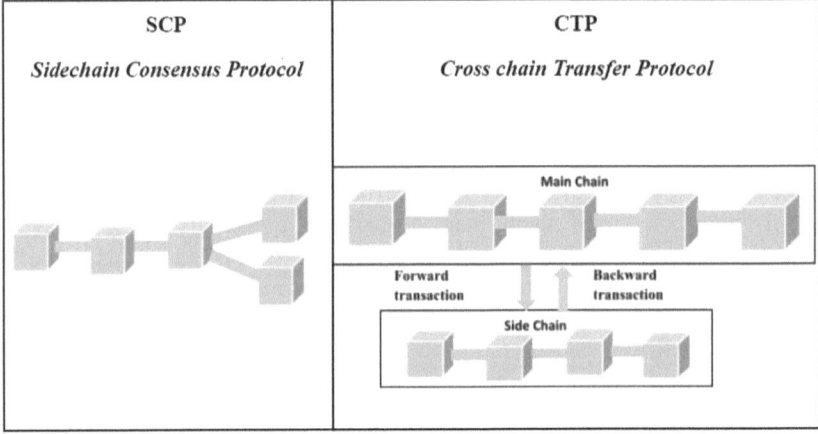

Figure 13.3 Side chain protocols

a trusted third-party validator to ensure smooth asset transfers between blockchains and are enforced and verified with low latency. A federated sidechain is one, when a third party is split into a group of validators with incentives to keep the system running. Each sidechain is only one to one with the mainchain, but we can build multiple sidechains. They are time-consuming and complex to initially build. However, once we build, any-body can interact with it. It is not completely trustless. We still need to rely on a trusted third party, the best being a federated sidechain [7].

The first project that utilized both relay and sidechain is the BTC Relay. This project works based on a technique called simplified payment verifica-tion (SPV), which verifies transactions cryptographically without the need to download the entire chain. The SPV client holds the chain of block headers displaying proof of work.

Sidechain technology is implemented in sample projects that include Elements, Loom, Mimblewimble, Poa network, Liquid, and RSK, and all the Polkadot bridges are basically sidechains [24].

The benefits of sidechain include increased scalability, serving as a testbed for new technologies, possibility of new interactions between new sidechains without intervention of the mainchain, and providing better security and privacy [26].

The drawbacks of sidechain include increased complexity due to the fact that sidechains are unsynchronized with the mainchain, even though being connected to it, and the possibility of fraudulent transfers and the lack of miners willing to participate in the coin mining operations.

13.4.2 Notary schemes

Notary schemes involve a notary, which is an entity that controls and monitors multiple blockchains. The notary initiates transactions in one chain, when an event takes place on another chain. Examples are Binance, Coin base, and Kraken. Basically, they are intermediaries like centralized exchanges. They are mostly centralized, but they offer speed and comfort over more decentralized solutions. They allow asset transfers between multiple different blockchains as shown in Figure 13.4.

Two categories of notaries exist [23]:

1. Custodians, who possess full control over a user's tokens locked in the smart contract vault (see Section 2.1.3) and are trusted to release those tokens whenever it is asked for.

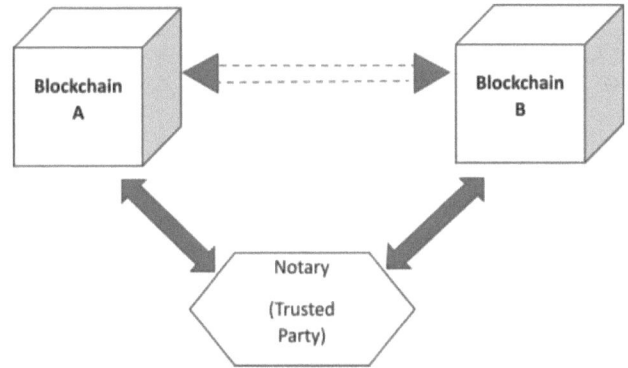

Figure 13.4 Notary schemes

2. External escrows, which possess only conditional control over a user's tokens locked in the smart contract vault. This is generally provided via a multi-signature contract in which both the signature of the user and the escrows are required before any transaction is being executed.

13.4.3 Hashed timelock contracts

Hashed time lock contracts allow transactions of assets in a trustless way. They are essentially programmable as escrows, and they are the technology behind atomic swaps. It allows fair exchange without a trusted third party, but it only works if the sender and the receiver provide their secret to each other and complete the transaction within the time lock. Both need to be online for this. Each hashed time lock contract must be set up individually for each use, for example, someone exploring the spread of crypto exchange rates, and they are limited to asset transfers. Some examples are decentralized exchanges and the lightning network [8].

Hashed time lock contract works on a transaction where two parties are involved. Both the parties publish a contract in which each party takes control of the asset possessed by the other party. The contract is agreed upon to store a pair (h, t). If it receives the corresponding hash lock secret 'a', $h = H(a)$, before the prescribed time 't' has passed, only then the contract

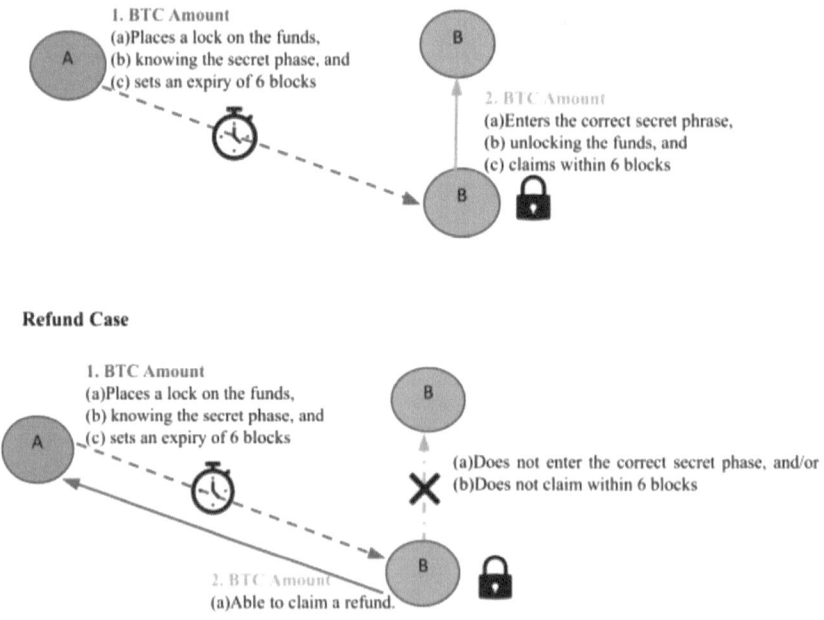

Figure 13.5 Hashed time-lock contracts

is successful and hence transfer of ownership of the asset to the counter-party is enabled. Otherwise, the asset is sent back to the original owner. Hash locking finds its usage mainly in atomic swaps. It also helps to incorporate transactions directed through several payment channels. Between parties, the time period within which each one must redeem their assets from the other varies across different chains. It works efficiently when sufficient time duration is provided between both the parties to redeem funds. Otherwise, the first party will deny to reveal the secret, and all funds will be returned. One of the projects that use hashed time lock is The Lightning Network [9].

Figure 13.5 depicts the successful case and the refund case of hashed time lock.

13.4.4 Blockchain of blockchains

Blockchain of blockchains is a framework that allows the creation of application-specific blockchains that can interoperate with each other. Therefore, they provide interoperability between blockchains based on the same architecture, for example, Polkadot, Kusama, Akala, Chain link, Ocean protocol, and Polymesh. However, it is still not interoperable in the truest sense. For example, to interoperate with EOS, a different blockchain, a mechanism similar to sidechains, still needs to be built. Currently, there is the EOS Bifrost Bridge. Some more examples of blockchain of blockchains are Ethereum 2.0, Cardano, Polkadot, Solana, Cosmos, and Algorand, as shown in Figure 13.6 [10].

13.4.5 Trusted relays

Trusted relays are entities that redirect transactions from one blockchain to another blockchain. They allow chains to confirm and verify events that have taken place in other chains. They work in private blockchains only.

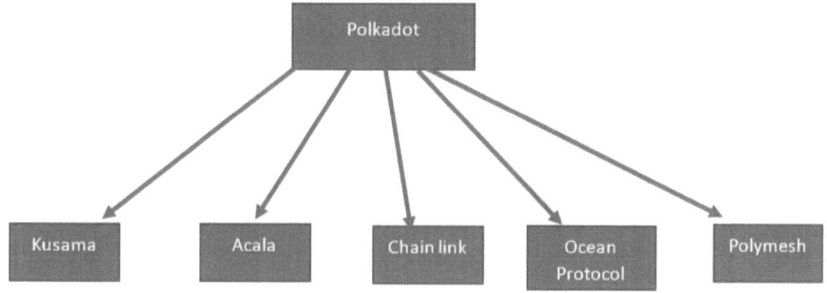

Figure 13.6 Blockchain of blockchains

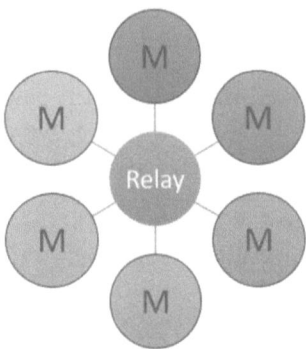

Figure 13.7 Trusted relays

Participants using trusted relays need to trust each other so that they can relay the information accurately. Therefore, they need to know of each other's identities beforehand. Most solutions are only theoretical at the moment, and an example would be Hyperledger cactus [27]. Figure 13.7 depicts how relays redirect transactions across blockchains.

13.4.6 Blockchain-agnostic protocols

Blockchain-agnostic protocols allow interoperability between multiple different blockchains by providing an abstraction layer. It is the necessity for the future of blockchains. The various blockchains can then settle by using the abstraction layer for accountability. The simplest way to understand what a blockchain-agnostic protocol does is that it acts as a kind of translator between blockchains. This works with both private and public blockchains. Some solutions lack flexibility to enforce smart contract logic on other blockchains. Some also lack flexibility to transfer non-fundable tokens. Some examples are the Interledger protocol, Hyperledger Quilt, and Quant Overledger. They still rely on trusted third parties. For the Interledger protocol, these parties are the connectors between the sender and recipient, whereas for Quant, it is the Quant Overledger and nodes on the system as shown in Figure 13.8 [11].

Mapps eliminate single-ledger dependency and allow companies to utilize the strengths of multiple blockchains without being restricted to their various limitations. Overledger's Multi-Chain Apps are very flexible. It helps in opening the door for unprecedented levels of business adoption [12].

Figure 13.9 depicts the current blockchain architecture and the role of Overledger in enabling Mapps.

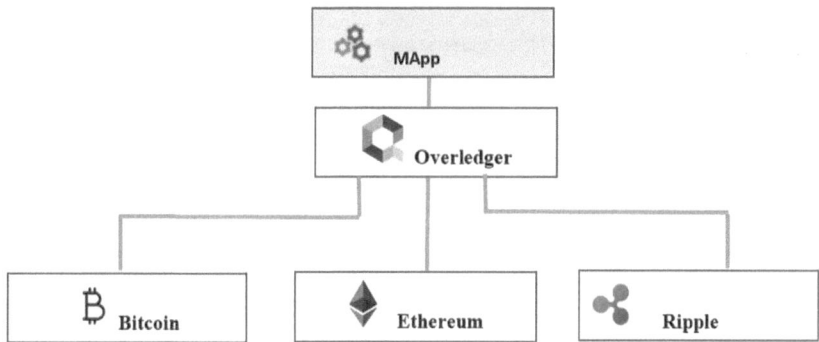

Figure 13.8 Multi-chain applications on Overledger.

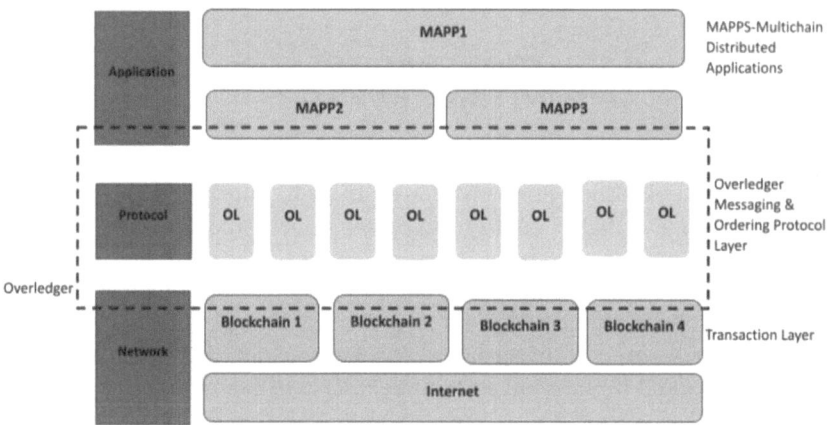

Figure 13.9 Current blockchain architecture and the role of Overledger in enabling MApps

13.4.7 Blockchain migrators

Blockchain migrators allow the migration of a complete state of a blockchain to another blockchain. These are specifically designed for migration in case of disaster or performance issues and are specifically designed for enterprise use cases. Most of the solutions are theoretical at the moment, especially migrativng smart contracts logic. One example is Hyperledger cactus [20].

Figure 13.10 depicts the working of blockchain migrators.

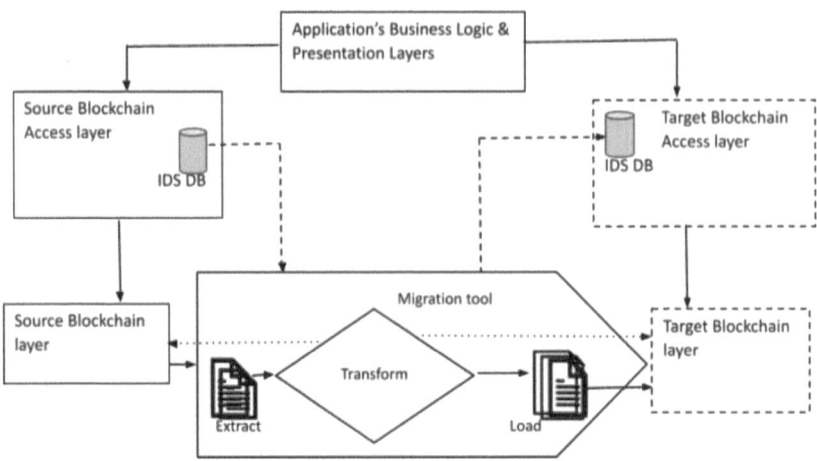

Figure 13.10 Blockchain migrators

13.5 EXISTING SOLUTIONS/PROJECTS OF BLOCKCHAIN INTEROPERABILITY

Some of the existing projects which implement blockchain interoperability are Cosmos, Polkadot, Harmony, Wanchain, Chainlink, Hybrix, and Loom network. Some of these solutions are open source and are easily available for all the networks.

Cosmos, the Switzerland-based company, offers the Cosmos SDK platform that supports blockchain interoperability across various systems and applications. Being a contract-based service provider, they offer blockchain solutions backed by optimal interoperability and scalability of blockchain systems [13].

Their blockchain architecture is based on a 'hub-and-spoke' system having a set of network 'spokes' communicating with the central hub to concert across multiple blockchain systems. Cosmos has defined a protocol "Inter-Blockchain Communication (IBC) protocol" based on this architecture. This protocol enables seamless IBC. It can also transfer tokens between Cosmos and other blockchains, which is compatible with the IBC protocol [14].

One of the goals of Cosmos is to establish an 'Internet of Blockchains (IoB)' which will act as the backbone for the communication between a multitude of applications in a decentralized manner [15].

Polkadot is a multi-chain technology initiative which involves multiple 'parachains' with distinctive characteristics. Having many chains interlinked in the network provides for long-distance transactions. This architecture also ensures high interoperability and robust data security. This blockchain enables high interoperability consistently across private chains, public networks, and others. Besides, this can scale up easily with less impact on governance.

Wanchain is another platform for developing blockchain solutions that are highly interoperable. The organization offers ultra-secure multi-party computing and blockchain solutions for the users. Interoperability and cross-chain capabilities are inherently supported by this blockchain leveraging the outcome of deep research anchored in cryptography. It also provides a proprietary protocol to connect clients of all types. Using Wanchain's cross-chain features, it is possible for any blockchain's assets to be connected with and circulate on Wanchain. The assets can be from public, private, or consortium chains.

Harmony is a blockchain-backed platform designed for decentralized applications to achieve scalability and security. Harmony has a cross-chain interoperability bridge called Horizon, which allows assets to be exchanged between the layer 1 Ethereum blockchain and the layer 2 Harmony blockchain. This solution allows Harmony to connect with the Binance blockchain, allowing the latter to reap the benefits of the former. Nodes on a different blockchain can validate transactions on the Harmony blockchain; thus, a much faster transaction speed can be achieved.

Blocknet is designed as a decentralized blockchain network supporting real-time interoperability based on vanilla implementation of inter-chain communication and data transfer. It proposes to 'define a new user perception to blockchain systems'. The unique and decentralized architecture matches the reliability of conventional blockchain tools. Organizations can benefit greatly from the cross-platform infrastructure by this blockchain network.

Blockchain interoperability can be achieved by using the cross-chain transaction technologies and the protocols mentioned above. It helps in accessing and sharing data between multiple blockchains by creating intermediate transactions or blockchains.

13.6 METHODS TO GENERATE INTEROPERABILITY IN BLOCKCHAIN NETWORKS

Blockchain interoperability can be a challenging task for developers to face. There exists two ways through which it can be achieved:

a. Mashed application programming interfaces (APIs) method
b. Network of networks method

13.6.1 API method

This mechanism achieves blockchain interoperability by transferring the data payloads with the help of application programming interfaces (APIs).

The 'mashed' API approach defines APIs to enable the communication between individual blockchain systems. The APIs have been designed with simple-to-use interfaces so that novice programmers can easily use them to enable blockchain interoperability.

Limitations:

1. Despite being flexible to use, the APIs suffer from the shortage of a clear governance structure. This makes it an unfavorable option for generating blockchain interoperability. Organizations will find it challenging to adhere to compliance regulations even if blockchain interoperability is achieved.
2. APIs will have to have a one-to-one mapping during integration with blockchain platforms leading to inefficient interoperability.
3. The involvement of APIs in the process could cause authentication deficiencies while data are transferred from one blockchain platform to another.

As a result of aforementioned constraints, the API method offers minimal interoperability in blockchain networks.

13.6.2 Network of networks method

The network of networks method is a superior way to build interoperability in blockchain networks as compared to the API method. The interoperability process can be made more efficient and scalable if this method is adopted. This method also establishes industry standards apart from allowing blockchain systems to identify if other blockchain networks are present to converge and communicate.

Since blockchain networks contain multiple network connections, the architecture can digitally connect blockchain networks and create interoperability possibilities across different applications. This process results in establishing blockchain network hubs for transmitting data through dedicated communication channels. This significantly brings down the complexity of blockchain interoperability.

The future looks quite promising with the advent of many methods for effective configuration of blockchain applications and networks.

13.7 OPEN PROTOCOLS

Use of open protocols is one of the best solutions for enhancing interoperability. They provide the universal language for blockchains, which enhances communication. Atomic swap is an open protocol which is used for blockchain interoperability. By acting as a decentralized escrow cross-chain, it allows the exchange of values between two different blockchains. While the transaction happens, there is no exchange or intermediary. Interledger uses the atomic swap protocol to implement the cross-chain blockchain.

13.8 MULTI-CHAIN FRAMEWORKS

Multi-chain frameworks behave like open environments. Different blockchains can be plugged into it, and hence, they become part of a standardized ecosystem. This facility allows sharing information among various blockchains. These frameworks offer distinct capabilities for blockchain interoperability. Due to this reason, they are also known as the IoB [16].

13.9 FEW USE CASES OF INTEROPERABLE BLOCKCHAINS

Many use cases demand interoperable blockchains as a catalyst for the delivery of digital services seamlessly from the stakeholders in private and public sectors. The lack of interoperable blockchains may lead to inefficient, costly, and fragmented service delivery. As per the paper "Blockchain Interoperability" published by the Technology and Innovation Lab of World Bank jointly with the Digital Advisory Unit of International Monetary Fund (IMF) in March 2021, organizations are uncertain about the advantage of blockchain technology for their operations and businesses; the primary impediment is the absence of interoperability.

It is quite evident that the transfer of information between different blockchains is a huge hurdle. Apparently, the organizations are wary of vendor lock-in which may eventually result in inability to transfer data if they wish to switch vendors.

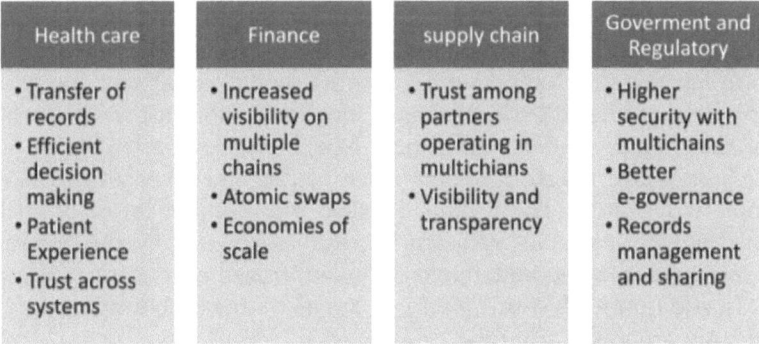

Figure 13.11 Blockchain usecases in different sectors

Figure 13.11 shows the few use cases of interoperable blockchains.

We all know that the value chain is very much essential in the *health care* sector. Therefore, blockchain interoperability can play a vital role in this sector. Interoperability guarantees an easy data flow within various blockchains. Using this facility, patients' and research data can be organized and maintained in an efficient manner. The current pandemic scenario requires the secure sharing of data for the government and various health institutions as well. An interoperable digital architecture is required for sharing the information.

Also, there is a need for interoperability in the financial-services sector as financial ecosystems hosted in heterogeneous blockchains can become a hindrance to financial entities and customers for the interactions, communications, and transactions. When the interoperability of blockchain becomes a reality, the transaction of data and value across financial ecosystems would be far more secure and cost-effective.

Practically, interoperability allows the constituent entities such as ecosystem users, developers, and applications to operate on the fabric built over numerous blockchain platforms. It allows a diverse spectrum of functionalities to be unleashed and encourages further innovation. When digital payment across several blockchains becomes a reality, the transfer of tokens and storage of digital assets from the convenience of just one wallet is going to be an exciting possibility. Such innovations will greatly enhance the overall user experience and lead to mass adoption of interconnected blockchains [17].

The supply chain industry is fragmented with every participant adopting his/her own approach. Blockchain definitely can introduce a unified standard and facilitate interoperability for seamless communication across various supply chain parties. There will be immense value added through the transparency ensured by interoperability by providing a source of truth for legitimate data garnered at various stages of the supply chain. This largely meets the customer expectations in terms of getting to know the holistic info about the products they purchase. Apart from delivering a plethora of data to the customers, blockchain can also help the supply chain participants to efficiently cooperate. The sheer possibility of deploying blockchain technologies to harness the true potential of future digital supply chain networks and platforms will create the necessary momentum for business growth.

For the supply chain domain, there have been many attempts to build blockchain platforms to overcome some of the inherent challenges. One example is Deloitte's TraceChain. Storing and tracking information like finished goods, materials, etc. can be easily supported by this platform. This information being immutable, an unauthorized party cannot alter the data. In addition to this, this platform can also draw rich insights into the production process in real time.

Despite the aforementioned value adds, blockchain will not solve all challenges faced by the supply chain sector. There is a high probability of the supply chain dealing with obsolete business processes and inaccurate product information. In case the source of particular ingredients itself is false or the information entered is inaccurate, such errors will remain persistently throughout the blockchain. In other words, blockchain like any other technology cannot compensate for the manual mistakes [18].

13.10 BENEFITS OF BLOCKCHAIN INTEROPERABILITY

Blockchain interoperability will have a key role in blockchain mining, cryptocurrencies, and trading markets.

Listed below are some of the benefits of blockchain interoperability:

- Sharing and integration of data among various blockchains.
- Enhancement of the capabilities of the emerging projects where the value chain is important. A few areas are healthcare, finance, trade, and aviation.
- Eased cross-chain transactions.
- More multi-token transactional operations.
- Extremely safe and secure transactions.

13.11 DISADVANTAGES OF BLOCKCHAIN INTEROPERABILITY

a. **It Is Not Possible to Reverse the Flow of Operations in Blockchain:** Before submitting a starting node, the data should be validated as the blockchain operations are irreversible.

b. **Sharing and Integration of Data between Two Different Networks Are Not Possible:** Currently, interoperability is possible for different blockchains of the same network. It is not possible for the two different blockchain networks, say, Ripple and Ethereum, to exchange their data with each other.

c. **Restricted Usage:** Interoperability ensures safe transfer of data, but still, it can behave like a complex and complicated feature for the users.

13.12 CHALLENGES IN BLOCKCHAIN INTEROPERABILITY

There exist several technologies, and these exist in silos of their own as well. There are pros and cons for each, and there is not a 'one size fits all' solution. The three main challenges in blockchain interoperability are as follows:

1. Almost all technologies still rely on a trusted third party. Therefore, if there was a new technology that allowed trustless third parties, that would be very interesting.
2. Another that goes unanswered is that all technologies are used for asset transfers, but what about smart contracts? There needs to be some ability to enforce a smart contract on one blockchain if a condition presents itself on another blockchain.
3. Also, the biggest challenge is the lack of standardization across the industry. There needs to be some agreed form of standardization across all blockchains in order for it to be true, scalable interoperability. For this to happen, there needs to be some global effort similar to how TCP/IP was maintained by global organizations like the IETF or the ISO. However, without this, interoperability is still achievable but with varying degrees of centralization and complexity.

Sidechains are most suitable for public blockchains, but creating and maintaining a decentralized application with multiple public and private blockchains becomes a challenge as we would need to build multiple custom sidechains. This is possible, but it takes time and effort.

Blockchain of blockchains solutions make this easier. However, blockchain of blockchains does not offer interoperability in the truest sense, for example, Polkadot – unless the ecosystem has a compelling selection of cryptos that fulfills decentralized applications needs, still one needs to use side chains. This does not mean that they are not good options. It is just that the current ecosystems are still very young and competition between them is fierce.

Hashed time lock contracts are also not interoperable in the truest sense. They are essentially contracts agreed upon by the sender and receiver, and so both entities must complete the transaction agreed upon the time frame. Therefore, value does not flow freely between networks here. This is more of a mechanism for trustless trade between two entities.

Notary schemes can be disregarded as they are mostly centralized. They are basically centralized exchanges.

Trusted relays are interesting but are mainly for private blockchains. For public blockchains, this would be a hard sell as most solutions have a central point of failure which is the entity doing the relaying. Unless there is some sort of trustless decentralized relay, this would not work for public blockchains.

Blockchain migrators are also mainly for private blockchains. The need for migration is useful for enterprise use cases. However, most solutions are theoretical.

This leaves blockchain-agnostic protocols to be the most viable solution for connecting both public and private blockchains. Blockchain-agnostic protocols are also easier to use compared to building sidechains. We know that technology always trends toward the most efficient and easy-to-use solution. Sidechains take time to build and are complex operations. If there was no side chain for a specific use case, then we probably use a blockchain-agnostic protocol first as it is easier. The overall conclusion is that blockchain-agnostic protocols are the easiest to use. Therefore, they will be the most popular. This is then followed by sidechains. Blockchain of blockchains is only good if we dig deep and do the due diligence.

13.13 CONCLUSION

Blockchain technology offers immense possibilities in different sectors such as healthcare and financial trade. In order to create ample output and profit from trade, interoperability of blockchains is very much essential. Interoperability helps in easing several intermediate processes and enhances database management procedures. It is an inevitable feature of blockchain to gain adoption across several industries. It enables crypto assets swapping across blockchains and allows sharing of supply chain records, health records, certificates, etc.

In recent times, numerous attempts have been made to integrate multiple blockchain systems to enable operational interoperability. Presently, these initiatives are limited to specific categories of blockchain networks in terms of scope and focus. However, the future looks quite promising with limitless possibilities for interoperability of blockchains. Once interoperability is established as the standard and norm in blockchain-based operations, the industry would witness rapid adoption of the technology. To achieve this, the innovation and research to enhance the interoperability aspect of blockchain applications and tools should be stimulated with large investments and focused efforts.

At present, blockchain interoperability is far behind the intended advancements due to the aforementioned challenges. Nonetheless, the cross-chain solutions resulting from recent research are indeed paving the way for better blockchain interoperability solutions in near future. In order to accelerate the overall pace, a number of enhancements are imperative, one such example being inter-network interoperability of blockchains. Innumerable benefits can be realized from blockchain interoperability upon tackling these challenges.

REFERENCES

1. Nakamoto, S. (2008). Bitcoin: A peer-to-peer electronic cash system. *Decentralized Business Review*, 21260. https://www.debr.io/article/21260. pdf

2. Soni, P. (2022). *All about Blockchain Interoperability in 2022*. https://www.analyticssteps.com/blogs/all-about-blockchain-interoperability-2022.
3. *A Guide on Blockchain Interoperability*. (2021). https://nownodes.io/blog/a-detailed-guide-on-blockchain-interoperability/.
4. Tse, S. (2021). *Blockchain Interoperability: Why It Matters and How to Make it Happen?* https://readwrite.com/blockchain-interoperability-why-it-matters-and-how-to-make-it-happen/.
5. Gaurav. (2021). *Blockchain Interoperability*. https://coincodecap.com/blockchain-interoperability.
6. Sidechains. (2019). https://academy.horizen.io/horizen/expert/sidechains/.
7. Kuznetsov B. (2017). *GitHub-Ethereum/Btcrelay: Ethereum Contract for Bitcoin SPV*. https://github.com/ethereum/btcrelay.
8. *Hash Time Locked Contracts-Bitcoin Wiki*. (2021). https://en.bitcoin.it/wiki/Hash_Time_Locked_Contracts.
9. Poon, J., Dryja, T. (2016). *The Bitcoin Lightning Network: Scalable Off-Chain Instant Payments*. https://lightning.network/lightning-network-paper.pdf.
10. ATB-Around the Block. (2021). *What Is Interoperability In Crypto And Best Projects To Invest In? [Video]*. YouTube. https://www.youtube.com/watch?v=8I99WORewD8&t=651s.
11. Potter, K. C. (2019). *Blockchain Agnosticism is the Future*. https://blockheadtechnologies.com/blockchain-agnostic-is-future/.
12. *Quant Overledger: The Solution to Single Ledger Dependency and the Promise of Multi-Chain Applications (MApps)*. (2018). https://medium.com/@djangoDLT/quant-overledger-the-solution-to-single-ledger-dependency-and-the-promise-of-multi-chain-38e25866b0a.
13. *Blockchain Interoperability: The Next Obstacle to Global Blockchain Adoption*. (2021). https://netfreeman.com/2021/10/20211004072126999k.html.
14. Takyar, A. *Blockchain Interoperability – Understanding Cross Chain Technology*, https://www.leewayhertz.com/blockchain-interoperability-crosschain-technology/.
15. Joshi, N. (2021). *Blockchain Interoperability: The Next Hurdle to Global Blockchain Adoption*. https://www.allerin.com/blog/blockchain-interoperability-the-next-hurdle-to-global-blockchain-adoption.
16. Geroni, D. (2021). *Blockchain Interoperability: Why Is Cross Chain Technology Important?*. https://101blockchains.com/blockchain-interoperability/.
17. Johnson, M. (2021). *Interoperability Is Fundamental to Blockchain's Future Technology*, https://internationalbanker.com/technology/interoperability-is-fundamental-to-blockchains-future/.
18. *Blockchain Interoperability in Supply Chain*. (2019). https://www.itransition.com/blog/blockchain-interoperability.
19. Back, A., Corallo, M., Dashjr, L., Friedenbach, M., Maxwell, G., Miller, A., et al. (2014). *Enabling Blockchain Innovations with Pegged Sidechains*.
20. Bandara, H.M., Xu, X., & Weber, I. (2019). *Patterns for Blockchain Migration*. ArXiv, abs/1906.00239.
21. Bodkhe, U., Tanwar, S., Parekh, K., Khanpara, P., Tyagi, S., Kumar, N., & Alazab, M. (2020). Blockchain for industry 4.0: A comprehensive review. *IEEE Access*, 8, 79764–79800. [9069885].

22. Huang, L.-Y., Cai, J.-F., Lee, T.-C., Weng, M.-H. (2020). A study on the development trends of the energy system with blockchain technology using patent analysis. *Sustainability.* 12, 2005. DOI: 10.3390/su12052005.

23. Lesavre, L., Varin, P., & Yaga, D. (2021). *Blockchain Networks: Token Design and Management Overview (No. NIST Internal or Interagency Report (NISTIR) 8301).* National Institute of Standards and Technology.

24. Mohanty, D., Anand, D., Aljahdali, H. M., & Villar, S. G. (2022). Blockchain interoperability: Towards a sustainable payment system. *Sustainability*, 14(2), 913. DOI: 10.3390/su14020913.

25. Mukherjee, P., & Pradhan, C. (2021). Blockchain 1.0 to blockchain 4.0 – The evolutionary transformation of blockchain technology. In *Blockchain Technology: Applications and Challenges* (pp. 29–49). Springer, Cham.

26. Musungate, B. N., Candan, B., Çabuk, U. C., & Dalkılıç, G. (2019). Sidechains: highlights and challenges. In *2019 Innovations in Intelligent Systems and Applications Conference (ASYU)* (pp. 1–5). IEEE.

27. Belchior, R., Vasconcelos, A., Guerreiro, S., & Correia, M. (2021). A survey on blockchain interoperability: Past, present, and future trends. *ACM Computing Surveys (CSUR)*, 54(8), 1–41.

28. Singh, A., Click, K., Parizi, R. M., Zhang, Q., Dehghantanha, A., & Choo, K. K. R. (2020). Sidechain technologies in blockchain networks: An examination and state-of-the-art review. *Journal of Network and Computer Applications*, 149, 102471.

29. Zamyatin, A., Harz, D., Lind, J., Panayiotou, P., Gervais, A., & Knottenbelt, W. (2019). Xclaim: Trustless, interoperable, cryptocurrency-backed assets. In *2019 IEEE Symposium on Security and Privacy (SP)* (pp. 193–210). IEEE.

Chapter 14

Blockchain for industry 4.0

Overview and foundation of the technology

Megha Gupta and Suhasini Verma
Manipal University Jaipur

CONTENTS

14.1 INTRODUCTION

Blockchain is one of the most significant technological breakthroughs in recent years. Blockchain is a decentralized database technology that has altered the way businesses operate. Companies and information technology (IT) giants have begun to spend considerably in the blockchain market, which is predicted to be valued at more than \$3 trillion in the next 5 years. It has grown in popularity due to its unrivaled security and capacity to provide a comprehensive answer to digital identification challenges. It is a peer-to-peer network's digital ledger (Sarmah, 2018). Blockchains are digital ledgers that resistant to tampering and are often implemented in a circulated structure (i.e. without a single repository) and without a central authority (i.e. a company, bank, or government. At its most basic, they permit a local area of clients to keep transactions in a shared ledger

DOI: 10.1201/9781003282914-14

inside that local area, so no transaction can be modified once recorded under standard blockchain network operation (Yaga et al. 2019).

The Oxford Dictionary describes blockchain as "a framework in which a record of Bitcoin or any other cryptocurrency transactions is maintained across numerous computers linked in a peer-to-peer network." However, the scope is not limited to cryptocurrencies, but the technology can be used in a variety of applications (Viriyasitavat and Hoonsopon, 2019). Blockchain is a distributed database of records or a public ledger of digital events or transactions that occurred and were shared among participating parties across a wide network of untrustworthy participants. It saves data in blocks that can validate data and are extremely tough to steal. It does away with the need for third-party verification, disrupting an industry that previously relied on it. Wherever a third party is involved in the production or a transaction, blockchain can be used to replace the third party. Each transaction in the public ledger is expected to be validated by the majority of system members, and once entered, information cannot be removed because it is immutable (Chakrabarti and Chaudhuri, 2017). It is utilized to get and convey information in a new and extraordinary manner. The absence of a central instance in a dispersed network implies a fundamental shift in how non-intermediaries or intermediary services interact with one another. Therefore, blockchain must be refreshed through an agreement among framework members, and an exchange can never be revised or deleted. Its circulated data set cannot be hacked, controlled, or intruded on like a norm, a centralized database with a client-controlled admittance conspires (Kitsantas, Vazakidis, and Chytis, 2019). To put it another way, when the information is mounted on the blockchain, nobody, can edit or remove anything from the ledger. Therefore, every information block is time-stamped and sequentially associated by utilizing a cryptographic mark (Walport 2016).

14.1.1 Overview of blockchain technology

These technology platforms have three generations: The Internet, mainframes, PCs, and local area networks, and the third platform, which conveys computing everywhere, instantly, and lets companies send and consume computing resources in shared networks. Blockchain technology is built on the third platform's capability (Efanov and Roschin, 2018). Blockchain technology is a progressive computer protocol that takes into account the advanced recording and capacity of data across various PCs or nodes. The purported "Ledger," which is analogous to a social data set, is one of the main components of blockchain (Kitsantas et al. 2019). Blockchain is a system that enables data storage and exchange on a peer-to-peer 1 (P2P) basis. Because of consensus-based algorithms, blockchain information may be shared, counseled, and secured structurally. It is utilized decentralized and eliminates the requirement for middle people or "trusted third parties." Blockchain arose from the union of two ideas:

1. Asymmetric cryptography, which employs a combined private and public key mechanism.
2. The IT architecture that is distributed (particularly P2P), PWC (2019).

The term "Blockchain" refers to two key ideas: (i) that all connected concurrent transactions are exclusively hashed and assembled into a "Block," which is cryptographically and interestingly distinguished by a Merkle root hash of every constituent hash, and (ii) every block is forever "Chained" to its immediate predecessor block. Merkle proofs, rather than the header's Merkle root, can be used to verify integrity throughout the whole blockchain (Ian and Emre, 2017). Unquestionably, the blockchain is another set of data sets. This innovation is particularly appealing to individuals since it can address one of the significant issues associated with finance. This is an issue of twofold spending without the use of a middleman (Golosova and Romanovs, 2018). Blockchain is an innovation that can cause enormous changes in our professional work and will have a critical effect over the course of the following, not many years. It can modify our view of business tasks and reshape our economy. It is a distributed and decentralized ledger system that, because it cannot be altered or faked, endeavors to guarantee data security, transparency, and last integrity. Blockchain is a public record framework that guarantees the respectability of transaction data. It is most popular as the technology that drives the Bitcoin cryptocurrency. When the Bitcoin cryptocurrency was introduced, blockchain technology was conveyed interestingly. Bitcoin is as yet the most broadly used implementation of blockchain technology to this day. Bitcoin is a decentralized digital currency payment system in view of the blockchain public transaction log. The important feature of Bitcoin is the currency's ability to keep its value in the absence of any organization or governmental authority in control. The Bitcoin network's number of transfers and users is constantly growing (Yli-Huumo et al., 2016).

Definitions of blockchain technology: "Blockchain is an incorruptible digital log of economic transactions that can be customized to record financial transactions as well as for all intents and purposes everything of significant worth" – this is one of the most prestigious implications of the blockchain, which was developed by Don and Alex Tapscott (Bahga and Madisetti, 2016).

As indicated by IBM, "Blockchain is a common, permanent record that makes it simpler to keep transactions and track assets in a corporate network." An asset can be either tangible (like a car, house, cash, or land) or intangible (patents, intellectual property, branding, copyrights). Almost everything of value may be recorded and traded on a blockchain network, cutting risk and costs for all parties.

14.1.2 Uses of blockchain technology in Industry 4.0

Germany coined the term "Industry 4.0" for year 2020 during the Hanover event in 2011. Industry 4.0 initiatives would result in dispersed, fully automated, and dynamic production networks powered by ten major technical enablers (Zhou, Liu, and Zhou, 2015). Enablers include the IoT, Internet, big data, blockchain, human–machine interaction, cloud computing and edge, artificial intelligence, robots, and open-source software. In Industry 4.0, the automation of industrial systems will be accomplished through interlinked (CPS) cyber-physical systems, permitting the modern foundation and creation cycles to change into an independent and dynamical framework. To work autonomously with one other and achieve the shared goal, the elements in this profoundly incorporated network should convey and function as smart devices. To support global economic, social, and environmental sustainability, (ICT) information and communication technologies are expected to play critical roles in sustainable industrialization (Weyer et al., 2015). Three central principles underpin Industry 4.0. The absolute first worldview is the smart product, which controls the resources and orchestrates the production process from start to finish. Traditional manufacturing processes are changed into dispersed, versatile, adaptable, and self-sorting production lines by the smart machine, which is a cyber-physical system. The final paradigm is the augmented operator, which augments a human operator's flexibility and capacity in an industrial system. Technologies will subsequently be used to assist the human operator, recognizing their importance when presented with a variety of tasks such as monitoring, specification, observing, and verifying creative processes. It strives to improve workers' capabilities and provide a collaborative workplace, reconfiguring their position in production cycles via human–machine interfaces that enable collaboration throughout the overall industrial ecosystem (Alladi et al., 2019).

14.2 STRUCTURE OF BLOCKCHAIN TECHNOLOGY

Blockchain is a conveyed information structure made up of a progression of blocks. Blockchain capacities are present as a distributed information base or worldwide record, keeping track of all transactions on a blockchain organization. The exchanges are time-stamped and grouped into blocks, every one of which is recognized by a cryptographic hash. The blocks are arranged in a linear order, with each block referencing the hash of the past block, producing a chain of blocks known as the blockchain. A blockchain

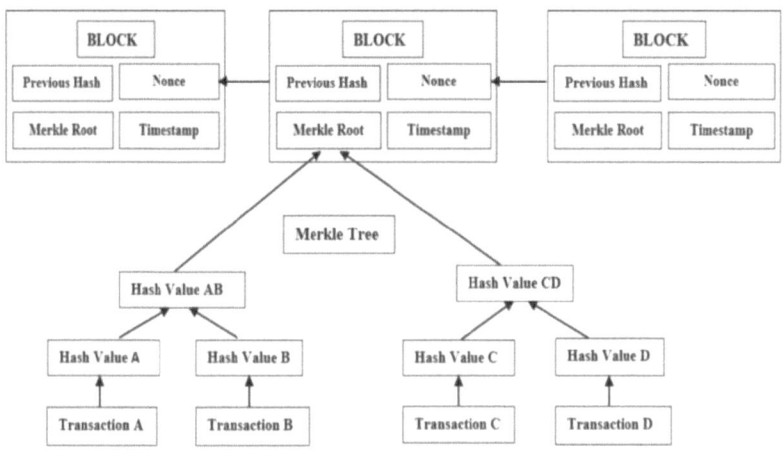

Figure 14.1 Construction of a blockchain.

is managed by an organization of nodes, each of which executes and records similar exchanges. The blockchain is duplicated among the blockchain organization's nodes. The exchanges can be perused by any node in the organization (Bahga and Madisetti, 2016). The construction of a blockchain is depicted in Figure 14.1.

In this structure, Hash means that when a transaction was finished, it was hashed to a code and then communicated to every node. Because every node's block may include a great many records of the transaction, blockchain uses the Merkle tree function to get a final hash value, along with the Merkle tree root, Merkle root means, the sum of the hash upsides of entire transactions in the block (Zheng et al., 2018), Nonce means, and a 4-byte field that ordinarily begins with 0 and ascends with every hash computation. Timestamp means the time when the block was created (Farah, 2018). The blockchain is arguably one of those technologies that, in terms of potential application domains, is now arousing high expectations. It is a global ledger that efficiently and permanently stores transactions on a chain of blocks. Each block comprises a collection of transactions that have been produced and dispatched throughout the system. Each block also has a timestamp, a connection to the past block, and is recognized by its hash value. Cryptographic hash capacities are utilized to sign and hash all transactions. Subsequently, this design gives an unforgeable log detailing the historical backdrop of all transactions made. The blockchain's nodes communicate with one another via a P2P network. Every node keeps a mirrored copy of the complete transaction history (Ferretti and D'Angelo, 2020).

14.2.1 How does blockchain work?

(Niranjanamurthy, Nithya, and Jagannatha, 2019). The first step is that the desired transaction is communicated to the P2P network, which is made up of computers known as nodes. The second step is validation: Using well-known techniques, the organization of nodes approves the transaction and the client's status. The third step is that a validated transaction might incorporate contracts, cryptocurrency, records, or different data. The fourth step is that after verification, the transaction is connected to other transactions to form another block of information for the ledger. The fifth step is that the new block is then for all time and irreversibly connected to the old blockchain. The sixth step is that the transaction has now been completed. Transactions are not legitimate till they are added to the chain. The tampering is readily visible. Because everyone in the network has a copy, the blockchain is considered secure. The sources of any differences are usually obvious right away (PWC Report 2019).

14.2.2 Why blockchain is important?

According to IBM, information is the soul of business. The quicker and more exact it is received, the better. Blockchain is astounding for conveying the data since it gives prompt, shareable, and totally straightforward information kept on a permanent ledger that must be viewed by network individuals with authorization. A blockchain organization can track installments, orders, records, production, and a variety of different things. What's more, since individuals have a brought-together viewpoint of reality, we can see all parts of a transaction from beginning to finish, offering us more prominent certainty as well as new efficiencies and amazing open doors.

14.2.3 Tiers of the blockchain

The blockchain technology tiers, that is, three, were initially depicted in Melaine Swan's book, Blockchain, Blueprint for a New Economy in light of the applications in every class (Figure 14.2) (Xu, Chen, and Kou, 2019; Efanov and Roschin, 2018; De Villiers and Cuffe, 2020; Yang, Garg, Raza, Herbert, and Kang, 2018).

  a. 1.0 Blockchain
     This blockchain is mostly utilized for cryptocurrencies, and it was first presented with the creation of Bitcoin. This blockchain tier includes all alternative currencies as well as Bitcoin. It also covers essential apps.
  b. 2.0 Blockchain
     This blockchain is used in a variety of financial services and businesses including options, financial assets, bonds, and, swamps, among

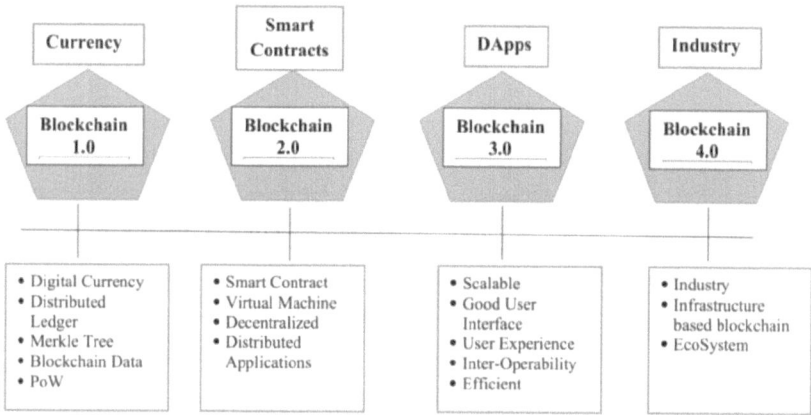

Figure 14.2 Evolution of blockchain technology.

other things. Smart contracts were originally presented in 2.0 block-chain and may be characterized as a method of verifying whether or not items and administrations are supplied by the provider during a transaction cycle between two gatherings.

c. 3.0 Blockchain

This provide greater safety than 1.0 and 2.0 Blockchain, and it is exceptionally versatile, adaptive, and sustainable. It is employed in a variety of areas, including health, art, justice, journalism, and numerous administration offices.

d. 4.0 Blockchain

This generation was fundamentally worried about ongoing services like public ledgers and conveyed databases. This level incorporates Industry 4.0-based applications seamlessly. It utilizes the smart contract, which eliminates the requirement for paper contracts and manages the network by agreement (Bodkhe, Tanwar, Parekh, Khanpara, Tyagi, Kumar, and Alazab, 2020; Fernandez-Carames and Fraga-Lamas, 2019).

14.3 THEORETICAL FOUNDATIONS OF BLOCKCHAIN TECHNOLOGY

In **1976**, a study titled "New Directions in Cryptography" was published, and it introduced the idea of a distributed ledger. With the evolution of cryptography, in **1991**, another study by Stuart Haber and Scott Stornetta titled "Hot to Time-Stamp a Digital Document" spread out the idea of timestamping the information rather than the medium. They upgraded their system in 1992 by using Merkle trees, which expanded productivity and permitted them to gather more documents on a solitary square. One more significant notion known as "Electronic cash" or "Digital Currency" that emerged in view of a model suggested by David Chaum additionally helped to the creation of the idea of blockchain, which was trailed by protocols, for example, e-cash conspires that presented twofold spending recognition (Sarmah, 2018; Popovski et al., 2018). Adam Back developed one more idea named "hash cash" in 1997, which suggested a way for spamming email control. Wei Dai developed the notion of "b-money" in view of a P2P network as a result of this. Satoshi Nakamoto is credited with inventing blockchain technology when he published "Bitcoin: A Peer-to-Peer Electronic Cash System" in 2008 (Nakamoto 2008; Zou, Meng, Zhang, Zhang, and Li 2020; Whitaker, 2019; Benton and Radziwill, 2017; Crosby et al., 2016), The abstract of the Nakamoto study focused on direct Internet-based payments starting with one source and then onto the next without the use of an outside source. The study depicted an electronic payment system in light of the cryptographic principle. Nakamoto's paper proposed an answer for twofold spending in which a digital currency could not be copied and could not be spent more than once. The paper presented the concept of a public ledger, where an electronic coin's transaction history can be traced and checked to guarantee that the coin has not been utilized already and to keep away from copy spending. The Bitcoin cryptocurrency blockchain network was established later in 2009. The blueprint that most modern cryptocurrency schemes follow is outlined in Nakamoto's paper (although with variations and modifications) (Yaga et al., 2019). Even though the creator of Bitcoins remained undisputed, Bitcoins were kept on being made and marketized, and a strong community was present to help and address different flaws with the code (Sarmah, 2018).

14.4 CHARACTERISTICS OF BLOCKCHAIN TECHNOLOGY

Blockchain technologies are made up of ten fundamental components, and these are as follows:

a. <u>Decentralization</u>: In traditional incorporated transaction frameworks, each transaction should be endorsed by a central trusted agency (e.g. the central bank), bringing about cost and execution inefficiencies at the central servers. Unlike the centralized solution, blockchain does not need a third party. Consensus algorithms are employed in blockchain to maintain data consistency in a distributed network (Naidu and Mishra, 2018). Because blockchain is decentralized, there is no need for a central institution, and every node is equal. It has evolved into the foundation technology of digital cryptocurrencies like Ethereum and Bitcoin (Li et al., 2020; Zou et al., 2020).

b. **Persistency:** Since each transaction disseminated across the network must be affirmed and kept in blocks scattered across the entire network, tampering is practically impossible. Furthermore, each communicated block would be approved by different nodes, and transactions would be checked. As a result, any deception might be quickly identified (Zheng et al., 2018).

c. **Anonymity:** The vital component of public blockchains is anonymity. This framework's ID can be untethered from a user's real-world identity. To avoid identity exposure, a single user can obtain many identities. No central entity is required to store private information (Yeow et al., 2017). Therefore, the real-world identity cannot be verified based on transaction information, preserving some privacy. In contrast, identity is typically required for frameworks that are worked and administered by known entities, such as private and permissioned blockchains (Viriyasitavat and Hoonsopon, 2019).

d. **Auditability:** Record timestamps and persistent metadata enable simple verification and tracing of earlier records across nodes in a blockchain network. The degree of auditability is determined by the type of blockchain technology and its implementation. Because nodes are managed by a single entity, private blockchains are the least auditable; permission blockchains are next in light of the fact that a few arrangements, like encrypted data, may prevent information from being completely auditable; and public blockchains are the most auditable on the grounds that nodes are truly decentralized (Viriyasitavat and Hoonsopon, 2019) (Figure 14.3).

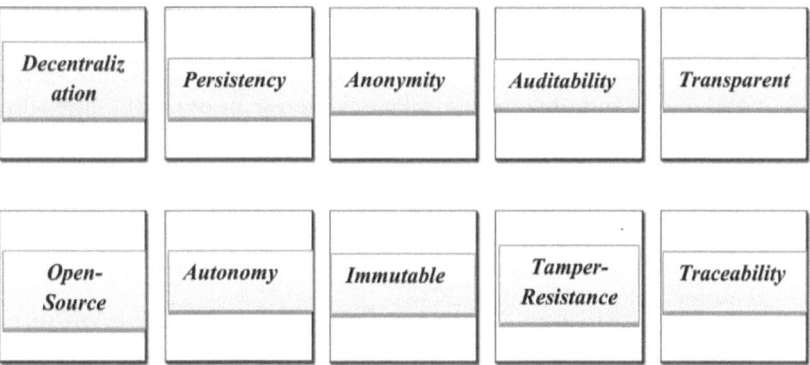

Figure 14.3 Characteristics of blockchain technology.

e. **Transparent:** The blockchain system's data record is straightforward to every node, and every one of these nodes can additionally refresh the information, making it transparent and trustworthy (Niranjanamurthy, Nithya, and Jagannatha, 2019).

f. **Open Source:** Most of the blockchain frameworks are available to the general population, records can be checked freely, and anyone can utilize blockchain technologies to construct any applications they desire.

g. **Autonomy:** On account of the agreement foundation, any node on the blockchain framework can safely move or update information; the objective is to trust from a solitary individual to the entire framework, and nobody can intercede.

h. **Immutable:** Any records will be saved in perpetuity and cannot be modified when somebody has control of in excess of 51% of the nodes simultaneously.

i. **Tamper Resistance:** Blockchain's tamper resistance implies that any transaction data saved in the blockchain cannot be altered during or after the block generation process. The blockchain's data structure is constructed by joining blocks carrying transaction information in an orderly fashion (Zou et al., 2020).

j. **Traceability:** Because of blockchain's traceability, transaction sources may be monitored via the data storage structure and chain structure. Li et al. explored fundamental advancements of the logistics information traceability model, while Xiao et al. devised a blockchain-based detectable IP copyright protection method (Li et al., 2020).

14.5 TYPES OF BLOCKCHAIN TECHNOLOGY

There are a few kinds of blockchains; the absolute most significant are (i) permissioned (public) blockchain, (ii) (hybrid blockchain) consortium blockchain, and (iii) non-permissioned (private) blockchain (Figure 14.4).

a. **Public Blockchain:** Public blockchains are accessible to the general population, and anyone can take an interest in dynamic interaction by turning it into a node; however, users may or may not benefit from their participation in the decision-making process. The ledgers are available to all network participants and are not owned by anyone on the network. The blockchain's users employ a distributed consensus process to reach a conclusion and keep a duplicate of the ledger on their local nodes (Sarmah, 2018).

b. **Consortium Blockchain:** This is a blend of private and public blockchain models. Using this architecture, organizations or institutions can create their own private blockchain network to distribute data among consortium members (such as institutions, banks, and other enterprises or firms) (Kitsantas, Vazakidis, and Chytis, 2019; Niranjanamurthy, Nithya, and Jagannatha, 2019).

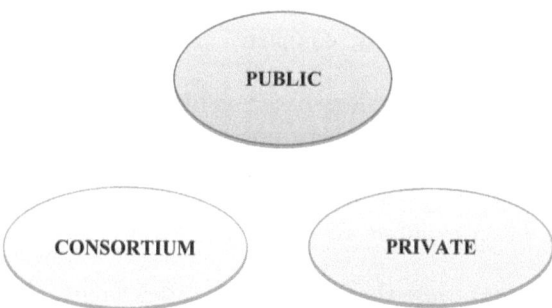

Figure 14.4 Types of blockchain technology.

c. **Private Blockchain:** The transactions are private, and the information is not accessible to people in general; however, the individuals are identified. A participant in a private blockchain network cannot access or compose the blockchain unless the participant is granted permission or is invited to join the organization. Private blockchain is typically utilized by major corporations, with permissions specified among the business Blockchain's numerous stakeholders. A bank, for example, can create its own blockchain network for private usage, with limited admittance to its many stakeholders like suppliers, clients, and workers (Kitsantas, Vazakidis, and Chytis, 2019).

14.6 BLOCKCHAIN TECHNOLOGY – ADVANTAGES AND DISADVANTAGES

This chapter highlights several advantages and downsides that must be addressed.

14.6.1 The blockchain technology – advantages

Dissemination is one of the primary advantages of blockchain, which permits a data set to be disseminated without the requirement for a focal power or element. On account of the blockchain's decentralized structure, it is practically hard to treat the information when contrasted with conventional data sets (Sarmah, 2018). Other advantages are as follows

a. **Data Integrity and Immutability:** Members can help to prevent fraud while also improving administrative consistency. When a record has been entered into the ledger, it must be erased with unanimous agreement.

b. **High Availability and Accessibility**: Blockchain technology information would be timely, complete, and exact due to decentralized networks.

c. **Reliability**: With technology based on blockchain, there is no one control center, and it can not be tempered.

d. **Decentralization**: Blockchain is a peer-to-peer transaction technology that eliminates the requirement for an outsider to function as an intermediary, hence avoiding all additional overhead costs and transaction fees.

e. **Transparency and Consensus**: All blockchain transactions are straightforward to any counterparty and take into consideration further audits at any moment. The shared ledger contains information on the destination, original source, date, and time of the transactions.

f. **Automation**: Blockchain technology makes use of smart contracts, where the terms of the contract may be saved and implemented on blockchain, which in turn, automate the process as per the terms of the contract.

g. **Processing Time**: Using blockchain technology, one may cut down the time it takes to process records or transactions from 3 days to seconds or minutes.

(Swan, 2015; Kitsantas, Vazakidis, and Chytis, 2019; Bahga and Madisetti, 2014; Christidis and Devetsikiotis, 2016).

14.6.2 The blockchain technology – disadvantages

If blockchain offers advantages, it also has drawbacks or issues. These are as follows:

a. **Cost Issues**: Blockchain technology entails high cost.

b. **Data Malleability Issues**: The pliability of data is a possible issue in blockchain deployment. The signatures do not constitute proof of ownership. An aggressor can change and rebroadcast a transaction, causing issues with transaction confirmation.

c. **Latency Issues**: The time factor is quite possibly the crucial difficulty in blockchain implementations since it is not suitable for large-scale transactions because of the rigorous verification procedure.

d. **Wasted Resources**: It necessitates a lot of energy. Mining in the Bitcoin network consumes around $15 million in electricity consistently.

e. **Integration Concerns**: Blockchain technology provides solutions that necessitate considerable adjustments to existing legacy systems in order to be implemented.

f. **Immaturity of the Technology:** Blockchain is another innovation that signifies a total transition to a decentralized network and may result in organizational change, including structure, changes in strategy, culture, and process.

(Kitsantas, Vazakidis, and Chytis, 2019; Beck, Stenum Czepluch, Nikolaj Lollike, and Malone, 2016; Swan, 2015; Aru, 2017; Yli-Huumo et al., 2016).

14.7 BLOCKCHAIN TECHNOLOGY – SWOT ANALYSIS

SWOT analysis (or SWOTM matrix) is a structured planning tool that assesses the four factors of strengths, weaknesses, opportunities, and threats (Niranjanamurthy et al. 2019). SWOT analysis can be used to assess new technologies or directions for a company or sector (Tezel et al., 2019). Blockchain, as a novel and innovative technology, necessitates strategic

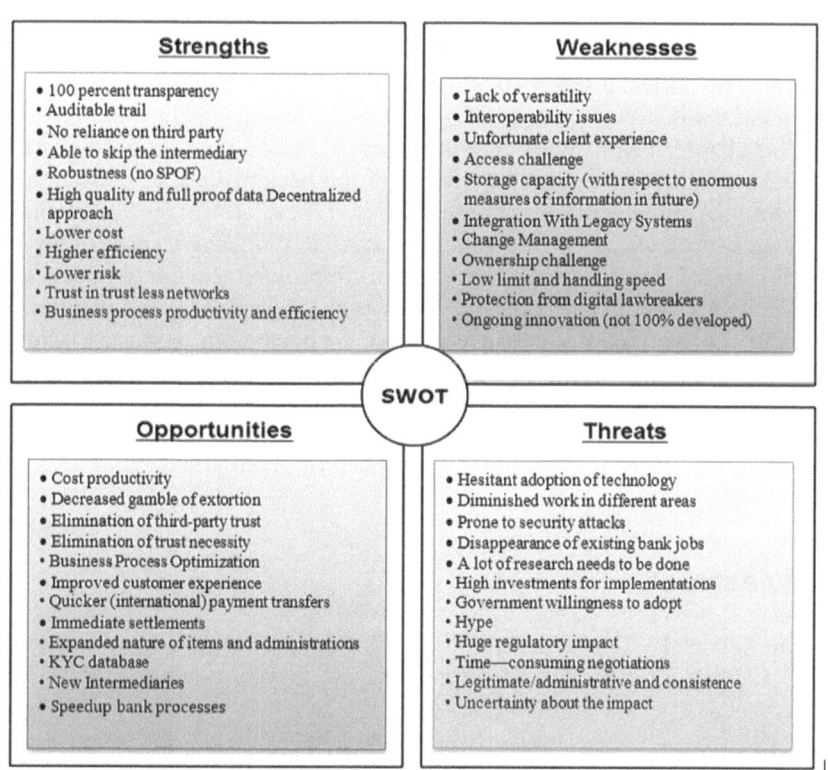

Figure 14.5 The blockchain technology – SWOT analysis.

planning study. To put it another way, before beginning to implement this technology in the forest business, it is important to conduct a SWOT analysis to recognize S (strengths), W (weaknesses), O (opportunities), and T (threats). S and W are internal environment factors, that is, those that the object can alter; O and T are environmental factors, those that can impact an object from the external perspective and are not constrained by the object (Vilkov and Tian, 2019). As a result, a SWOT analysis was performed, as shown in Figure 14.5.

14.8 CONCLUSION

Blockchain technology is beneficial and flexible in our world since it can simplify the majority of systems in many industries. According to a survey of the literature, blockchain technology has great theoretical value and promising possibilities for resolving data integrity issues, boosting transparency, enhancing security, establishing trust and privacy, and reducing fraud. In recent years, blockchain technology research has been actively conducted at numerous universities, with the amount and nature of published papers improving further. Therefore, a slew of blockchain initiatives in a variety of sectors has emerged, with a wide scope of application situations that can be coupled with the Internet of Things, medical, supply chain, and other different fields.

This chapter begins with defining blockchain and then proposes a structure for blockchain technology based on the background. It next explores the foundation, overview, characteristics, types, advantages, and disadvantages of blockchains in business processes. Blockchain technology has demonstrated tremendous potential for improving corporate processes. Blockchain's qualities of validity, persistency, auditability, and disintermediation can significantly improve modern corporate operations in terms of automation, digitalization, and transparency. According to the findings of this study, it is possible to conclude that blockchain aids in removing the involvement of third parties in any transaction. It can be used to assist in avoiding fraud and forgeries in a range of businesses.

REFERENCES

Adam-Kalfon, P., El Moutaouakil, S., & Richard, C. (2017). *Blockchain, Is a Catalyst for New Approaches in Insurance*. Paris: PwC. https://www.PWC.com/gx/en/insurance/assets/blockchain-a-catalyst.pdf, Erişim Tarihi, 6, 2019.

Alladi, T., Chamola, V., Parizi, R. M., & Choo, K. K. R. (2019). Blockchain applications for industry 4.0 and industrial IoT: A review. *IEEE Access*, 7, 176935–176951.

Aru, I. (2017). Full stack development tools lowering blockchain entry. *News Cointelegraph*. Available at: https://cointelegraph.com/news/full-stack-development-tools-lowering-blockchain-entry-barriers.

Bahga, A., & Madisetti, V. (2014). *Internet of Things: A Hands-on Approach*. VPT.

Bahga, A., & Madisetti, V. K. (2016). Blockchain platform for industrial internet of things. *Journal of Software Engineering and Applications*, 9(10), 533–546.

Beck, R., Stenum Czepluch, J., Nikolaj Lollike, N., & Malone, S., (2016). Blockchain-the gateway to trust-free cryptographic transactions. In *ECIS. Research Paper 153*.

Benton, M. C., & Radziwill, N. M. (2017). Quality and Innovation with blockchain technology. arXiv preprint arXiv:1710.04130.

Bodkhe, U., Tanwar, S., Parekh, K., Khanpara, P., Tyagi, S., Kumar, N., & Alazab, M. (2020). Blockchain for industry 4.0: A comprehensive review. *IEEE Access*, 8, 79764–79800.

Chakrabarti, A., & Chaudhuri, A. K. (2017). Blockchain and its scope in retail. *International Research Journal of Engineering and Technology*, 4(7), 3053–3056.

Christidis, K. & Devetsikiotis, M. (2016). Blockchains and smart contracts for the internet of things. *IEEE Access*, 4, 2292–2303.

Crosby, M., Pattanayak, P., Verma, S., & Kalyanaraman, V. (2016). Blockchain technology: Beyond bitcoin. *Applied Innovation*, 2(6–10), 71.

De Villiers, A., & Cuffe, P. (2020). A three-tier framework for understanding disruption trajectories for blockchain in the electricity industry. *IEEE Access*, 8, 65670–65682.

Efanov, D., & Roschin, P. (2018). The all-pervasiveness of the blockchain technology. *Procedia Computer Science*, 123, 116–121.

Farah, N. A. A. (2018). Blockchain technology: Classification, opportunities, and challenges. *International Research Journal of Engineering and Technology*, 5(5), 3423–3426.

Fernandez-Carames, T. M., & Fraga-Lamas, P. (2019). A review on the application of blockchain to the next generation of cybersecure industry 4.0 smart factories. *IEEE Access*, 7, 45201–45218.

Ferretti, S., & D'Angelo, G. (2020). On the Ethereum blockchain structure: A complex networks theory perspective. *Concurrency and Computation: Practice and Experience*, 32(12), e5493.

Gatteschi, V., Lamberti, F., Demartini, C., Pranteda, C., & Santamaría, V. (2018). Blockchain and smart contracts for insurance: Is the technology mature enough? *Future Internet*, 10(2), 20.

Golosova, J., & Romanovs, A. (2018). The advantages and disadvantages of the blockchain technology. In *2018 IEEE 6th Workshop on Advances in Information, Electronic and Electrical Engineering (AIEEE)* (pp. 1–6). IEEE.

Ian, P., & Emre, E. (2017). *Perspectives of Blockchain Technology, its Relation to the Cloud and its Potential Role in Computer Science Education*.

Junejo, A. Z., Memon, M. M., Junejo, M. A., Talpur, S., & Memon, R. M. (2020). Blockchains technology analysis: Applications, current trends and future directions – an overview. *Intelligent Computing and Innovation on Data Science*, 411–419.

Kitsantas, T., Vazakidis, A., & Chytis, E. (2019). A review of blockchain technology and its applications in the business environment. In *International Conference on Enterprise, Systems, Accounting, Logistics & Management, Chania, Crete, Greece, July 2019.*

Li, X., Lv, F., Xiang, F., Sun, Z., & Sun, Z. (2020). Research on key technologies of logistics information traceability model based on consortium chain. *IEEE Access*, 8, 69754–69762.

Li, X., Mei, Y., Gong, J., Xiang, F., & Sun, Z. (2020). A blockchain privacy protection scheme based on ring signature. *IEEE Access*, 8, 76765–76772.

Naidu, G., Mishra, R. (2018). Blockchain technology architecture and key characteristics. IJARIIE, 4(4), ISSN (O)-2395-4396.

Nakamoto, S. (2008). Bitcoin: A peer-to-peer electronic cash system. *Decentralized Business Review*, 21260.

Popovski, L., Soussou, G., & Webb, P. B. (2018). A brief history of blockchain. Online access: February 1, 2019.

Weyer, S., Schmitt, M., Ohmer, M. & Gorecky, D. (2015). Towards industry 4.0-standardization as the crucial challenge for highly modular, multi-vendor production systems. *IFAC-Papers Online*, 48(3), 579–584.

Sarmah, S. S. (2018). Understanding blockchain technology. *Computer Science and Engineering*, 8(2), 23–29.

Swan, M. (2014). *Blockchain-Enforced Friendly AI. Crypto Money Expo.* http://cryptomoneyexpo.com/expos/inv2/#schedule   and   http://youtu.be/qdGoRep5iT0/.

Swan, M. (2015). *Blockchain: Blueprint for a New Economy.* O'Reilly Media, Inc.

Tezel, A., Papadonikolaki, E., Yitmen, I., & Hilletofth, P. (2019). Preparing construction supply chains for blockchain: An exploratory analysis. In *Proceedings of the CIB World Building Congress' Construting Smart Cities'.* International Council for Research and Innovation in Building and Construction (CIB).

Vilkov, A., & Tian, G. (2019). Blockchain as a solution to the problem of illegal timber trade between Russia and China: SWOT analysis. *International Forestry Review*, 21(3), 385–400.

Viriyasitavat, W., & Hoonsopon, D. (2019). Blockchain characteristics and consensus in modern business processes. *Journal of Industrial Information Integration*, 13, 32–39.

Whitaker, A. (2019). Art and blockchain: A primer, history, and taxonomy of blockchain use cases in the arts. *Artivate*, 8(2), 21–46.

Xu, M., Chen, X., & Kou, G. (2019). A systematic review of blockchain. *Financial Innovation*, 5(1), 1–14.

Yaga, D., Mell, P., Roby, N., & Scarfone, K. (2019). *Blockchain Technology Overview.* arXiv preprint arXiv:1906.11078.

Yang, W., Garg, S., Raza, A., Herbert, D., & Kang, B. (2018). Blockchain: Trends and future. In *Pacific Rim Knowledge Acquisition Workshop* (pp. 201–210). Springer, Cham.

Yeow, K., Gani, A., Ahmad, R. W., Rodrigues, J. J., & Ko, K. (2017). Decentralized consensus for edge-centric internet of things: A review, taxonomy, and research issues. *IEEE Access*, 6, 1513–1524.

Yli-Huumo, J. et al., (2016). Where is current research on Blockchain technology? A systematic review. *PLoS One*, 11(10), e0163477. DOI: 10.1371/journal. pone.0163477.

Yli-Huumo, J., Ko, D., Choi, S., Park, S., & Smolander, K. (2016). Where is current research on blockchain technology? – A systematic review. *PLoS One*, 11(10), e0163477.

Zheng, Z., Xie, S., Dai, H. N., Chen, X., & Wang, H. (2018). Blockchain challenges and opportunities: A survey. *International Journal of Web and Grid Services*, 14(4), 352–375.

Zhou, K., T. Liu, and L. Zhou. (2015). Industry 4.0: Towards future industrial opportunities and challenges. In *Proceeding of 12th International Conference Fuzzy System Knowledge Discovery (FSKD)*, pp. 2147–2152.

Zou, Y., Meng, T., Zhang, P., Zhang, W., & Li, H. (2020). Focus on blockchain: A comprehensive survey on academic and application. *IEEE Access*, 8, 187182–187201.

Chapter 15

Enabling cross-border trade in the face of regulatory barriers to data flow – the case of the blockchain-based service network

Shoufeng Cao, Xavier Boyen, Felicity Deane,
Thomas Miller, and Marcus Foth
Queensland University of Technology

CONTENTS

15.1 INTRODUCTION

The adoption of digital technologies to facilitate international trade, coupled with the acceleration of digital transformation in the movements of goods and services, has led to exponential growth in data exchanges across countries (Meltzer and Lovelock, 2018). According to UNCTAD (2021), the world has experienced exponential growth in the volume and velocity of the flow of digital data across borders. The growth in cross-border data flows has coincided with the growth in cross-border trade and is expected to outperform the growth of global trade (Cory, 2017), especially after the coronavirus disease 2019 (COVID-19) pandemic, which has acted as an accelerating catalyst for developing the digital economy worldwide (Kumar, 2020). Alongside the growing data exchanges across national borders, increasing data availability, access, and sharing to maximize the economic and social value to companies, consumers, and national economies have been widely recognized (Nguyen and Paczos, 2020; Snelson and Cilauro, 2021).

Despite positive data-driven value and trade growth, the regulations that govern cross-border data flows specific to each trading country's jurisdictions are increasing, which could slow down cross-border data flows (Cory and Dascoli, 2021). Regulations are vital to protect producers, consumers, and the data sovereignty of nation states (Foth et al., 2021; Herian, 2020; Mann et al., 2020; Thumfart, 2022); the slower pace of regulatory reform work could consequently impair the value growth potential of data and act as a brake on future data-dependent trade growth (Cory, 2017). Proponents of the neoliberal "free market" mantra support this and suggest that the flow of data – whether analog or digital – is a precondition for moving goods and services across borders (Steger and Roy, 2010). They often label any regulation as "red tape" that has to be cut and removed. We caution against this argument given that there is little empirical evidence to suggest that the introduction of data flow restrictive legislation and regulations has curtailed the growth in trade in goods and services. If the 'free flow of data' is increasingly hampered by lagging regulatory governance, this can risk creating bottlenecks. However, we argue that a workable solution is not achieved by the mere removal of regulation but by investing in regulatory reform work that is not limited by technological solutionism (Herian, 2020; Morozov, 2014).

This chapter is exploratory in nature seeking to examine how regulatory and technical responses – blockchain-related technologies in this case – can ensure compliance with cross-border rules and regulations for data flows. There are several substantial initiatives emanating from China that seek to address the challenges of cross-chain data flows in an across-jurisdiction data exchange context. We examine how some initiatives in China, insofar as data regulations are concerned, are seeking to address challenges given rise to by new regulatory systems of governance and establish a jurisdictionally compliant digital environment that can enable cross-border trade in an increasingly regulatory landscape. This acts as something of an 'ideal type' (in Weber's sense), enabling us to explore an extensive example of data regulation and how trade nonetheless can proceed regardless. In other words, we seek to explore the question: how can digitally enabled cross-border transactions take place when regulations in one or many jurisdictions potentially curtail the speed of relevant data movements? We do so by exploring one practical initiative – the Blockchain-based Services Network (BSN) – emanating from China, which seeks to establish a jurisdictionally compliant digital environment that can enable cross-border trade in conditions of constrained cross-border data flows. In exploring this case study, the aim of this chapter is to neither endorse nor dismiss the work but to use it to open up a better understanding of the nature of the challenges inherent in policy reform work and the associated architectural and/or technical solutions that can be envisaged to accompany – rather than replace – it.

By linking blockchain-enabled infrastructure and cross-border data flows in the face of rising regulatory barriers within an international trade context, this chapter makes both technical and practical contributions in several ways. First, it explores whether technical measures could be a potential solution to enable cross-border trade with undisrupted flows of data across national borders while meeting the rising regulatory requirements to govern cross-border data flows in a trans-national context. Second, it analyzes the emerging BSN as a technical instrument for facilitating global cross-border data flows within regulatory governance systems in force between countries. It also reflects on the limitations of the BSN approach associated with its architectural design for cross-border trade in accordance with national/trans-national regulatory requirements. Third, this chapter proposes a BSN-based 'hub-to-hub' cross-border trade framework that incorporates Zero Knowledge Proofs (ZKPs) into multi-party governance mechanisms in a trans-national trade context and raises directions for future study. In a broad sense, this chapter informs scholars and professionals of how blockchain-based infrastructure can be used to transform and govern cross-border trade in the rising regulatory landscape. It thus aims to advance the application of blockchain technology for industry transformations that meet the needs of trans-national data governance and sovereignty concerns.

The chapter is organized as follows: Section 15.2 reviews the relationship between cross-border data flows and international trade. Section 15.3 presents an overview of the evolving regulatory interventions on the flow of data across borders. Section 15.4 discusses some of the measures and instruments that have been introduced to enable cross-border data flows. This is followed by the introduction and description of the BSN in Section 15.5. Section 15.6 reflects on the BSN's technical approach and its limitations for cross-chain and cross-jurisdiction data exchange for international trade and uses it to raise some further considerations relevant to the design of architectures that can enable cross-border commerce to grow despite the increasing regulatory barriers to cross-border data flows. This chapter concludes with our comments on the BSN from a technical point of review and our solutions for improving BSN's capability for cross-border data flows, research limitations, and future directions in Section 15.7.

15.2 CROSS-BORDER DATA FLOWS AND INTERNATIONAL TRADE

Cross-border data flows refer to 'the movement of data across borders between two or more computer servers' (Laidlaw, 2021). The global spread of the Internet and the rise of disruptive technologies have facilitated such information sharing across borders, which is increasingly shaping the world's trade and economy (Martinez, 2021; Meltzer, 2015b). Since late 2019, the current COVID-19 pandemic has acted as a catalyst for unprecedented growth in digital commerce and trade. As noted by Gallaher and Cory (2020), since the outbreak of the global pandemic, cross-border e-commerce has experienced exponential growth globally. For instance, e-commerce orders in the United States and Canada grew by 129% from 2019 to 2020. The growth in cross-border e-commerce has accelerated the digital transformation of international trade in goods and services (Lee-Makiyama, 2018), which results in a steady increase in the flow of data across borders.

The finalization of trade transactions needs the collection and use of data and digital information, including personal information such as names, addresses, and billing information. Data have therefore become a vital component of international trade in the digital era (National Board of Trade, 2014). Pepper et al. (2016) found that free data flows across borders reduce costs associated with trade and transactions, including customer acquisition and order fulfilment. Mitchell and Hepburn (2018) concluded that data transfer across borders supports trade in both service industries and traditional industries. Meltzer and Lovelock (2018) and Casalini and González (2019) argued that cross-border data flows transform the processes of international trade and help firms expand to international markets and adapt their products and services to the market demand. Each of these commentators presents a unique argument that recognizes the importance of cross-border data flows to the global economy. This development is further expedited by a parallel trend in product-related consumer communication data being exchanged on social media and digital platforms (Cao et al., 2021a; Choi et al., 2014; Hearn et al., 2014).

Companies, consumers, and national economies can benefit from the availability of data and easy access and sharing data across borders in international trade (Castro and McQuinn, 2015; Cory, 2017). This is echoed by Snelson and Cilauro (2021) and Nguyen and Paczos (2020), who argued that the availability of data can create numerous direct and indirect economic and social benefits to businesses and economies. This evidence supports that streamlining cross-border data flows can enhance direct and indirect economic and social value of data (Nguyen and Paczos, 2020), while data security and privacy are critical to ensure consumer and business confidence, enhance trust, and minimize cybercrimes. This requires a balanced solution for streamlining data flows across borders, to enable free trade in the digital age, while acting as a key enabler for safe and reliable international trade in the digital age. This chapter responds to this research need by identifying a fit-for-purpose solution.

15.3 REGULATORY GOVERNANCE OF CROSS-BORDER DATA FLOWS

In the previous section, we have demonstrated that international trade requires data flows between countries. Despite recognizing that cross-border data flow is essential to international trade, there are increasing concerns about protecting privacy, data flow governance, and trust (Kay, 2019). Laidlaw (2021) identified three overlapping challenges: (i) data security or cybersecurity challenges; (ii) privacy challenges, and (iii) information challenges pertaining to misinformation or sensitive data. In view of these challenges, there have been several proposals for regulatory reform work, including the advancement of consumer protection in online environments (Meese et al., 2019), interoperability standards (Keogh et al., 2020), protection of digital assets (Valeonti et al., 2021), and promotion of open source technology (Mitchell and Mishra, 2018). However, due to the variation of legal and regulatory interventions relevant to data flow governance across countries, data flow regulation across borders is challenging (Mitchell and Mishra, 2019), which has resulted in what have been referred to as regulatory barriers, at regional and international levels.

Several countries have implemented laws to impede the cross-border exchange of data and promote the pursuit of, for example, national security, public morals or public order, data sovereignty, and privacy (Bünz et al., 2020; Couture and Toupin, 2019; Gruin, 2021; Mitchell and Mishra, 2018; Thumfart, 2022). Data localization measures that restrict the flow of data from one country/jurisdiction to another are also increasingly on the rise in the past 4 years. According to Cory and Dascoli (2021), 144 data localization barriers have been implemented by 62 countries in 2022 compared with 67 such barriers by 35 countries in 2017. These policies and measures are generally designed for data integrity, cybersecurity, and online consumer protections. However, these regulations come with a cost (Mitchell and Mishra, 2019). Bauer et al. (2013) estimated that disruptions to cross-border data flows and service trade could cause the European Union's gross domestic product (GDP) loss of up to 1.3% and a potential drop of up to 11% in their manufacturing exports to the United States. The United States International Trade Commission (2014) estimated that the US GDP would increase up to 0.3% by decreasing barriers to cross-border data flows. These estimates highlight that the essential consumer protections and benefits afforded by data regulations come with a price tag that has to be factored into the economic equation of the global economy.

Despite these estimates, in the European Union, the GDPR was enacted in 2018[1] under the objective that it provides protection for the fundamental right to the protection of personal data. The protection offered by the GDPR comes with a barrier insofar as it requires personal information to remain in the territory, prohibiting the transfer of data to a third-party nation unless an exemption applies. In addition, China has recently introduced a collection of cybersecurity and data privacy regulations. In November 2016, China passed its first Cybersecurity Law, and in 2021, the Data Security Law (DSL) and the Personal Information Protection Law (PIP) added to the

---

[1] https://www.gov.uk/data-protection

data protection framework.[2] The DSL and PIP reflect some features of the GDPR, despite the policy differences that underpin them. One of the main points of difference between the two jurisdictions is the prevalence of data sovereignty and national security that underpins the Chinese legislation. Where the European Union has attempted to promote the policy objective of personal privacy (with what may or may not be unintended consequences of data localization), the Chinese legislation explicitly allows the state to access data in a host of circumstances. As such, there is the added challenge of reconciling policy objectives alongside finding a balance to support fair and open digital trade. This chapter makes efforts to respond to this challenge from both technical and practical perspectives and propose suitable instruments to facilitate free data flows across national borders.

15.4 INSTRUMENTS FOR ENABLING CROSS-BORDER DATA FLOWS

Concerns about achieving a more balanced approach for regulatory governance of cross-border data flows as discussed in the previous section have led to (i) the use of existing trade rules and (ii) the development of new instruments. The World Trade Organization (WTO) trade agreements offer some guidelines on data transfer across nations (Meltzer, 2015b). However, these rules were developed in the pre-Internet era and thus have limited applications in addressing the emerging issues of cross-border data flows in the digital era (Mitchell and Mishra, 2018, 2019). Although there are some strengths in negotiating new rules through the forum of the WTO, new trade rules for cross-border data flows at national, regional, and international levels are necessary to support international trade while allowing for data sovereignty where it is desired. More recently, the Comprehensive Trans-Pacific Partnership (CPTPP) and the Regional Comprehensive Economic Partnership (RCEP) have included specific articles that prohibit data localization requirements. Both these agreements include exception provisions, which differ slightly but generally require the desire to achieve a legitimate public policy objective.

These trade agreements and regional regulatory frameworks help address some of the concerns about data privacy and consumer protection related to cross-border data flows, while providing a balanced approach to support specific policy objectives. However, these rules leave substantial gaps and do not specifically address domestic concerns in relation to privacy, cybersecurity, and data sovereignty (Martinez, 2021). Further, the exception provisions are often extensive, causing one to question the signatory parties' commitments to open cross-border data flow. Although these regulatory issues must also be separately addressed, here we argue for the development

---

[2] https://www.lexology.com/library/detail.aspx?g=92d06cc9-22e3-44e6-b28e-860c25ef7630

of complementary practical instruments to support digitally enabled trade relationships between nations as a hybrid approach that combines the best of both worlds: technology and policy.

Technical measures are regarded as a practical – yet limited on their own – instruments for addressing safety and security associated with cross-border data flows in the digital era (Herian, 2020). As noted by Casalini and González (2019), technical factors, including encryption and cybersecurity, play a critical role in ensuring data safety and security. However, interoperable standards and policies are essential for any technical instrument or measure to be successful when deployed on data flows across borders (Keogh et al., 2020; Meltzer, 2015a). As a result, some countries have worked together to negotiate and develop consensual digital technical standards and policies. In May 2021, G7, together with Australia and the Republic of Korea, agreed on a Ministerial Declaration to develop a collaborative framework for digital technical standards, which includes a focus on electronically transferable records to promote digital trade by reforming regulatory legacy systems (Roxas, 2021).

Blockchain was invented as a distributed and immutable ledger for data sharing (Dedeoglu et al., 2020; Foth, 2017) and has evolved as a technical solution for tackling interoperability issues. Further, blockchain uses public-key/asymmetric cryptography and hash functions to maintain on-chain transactions in the form of immutable records (Fernández-Caramès and Fraga-Lamas, 2020), thereby ensuring secure and reliable data sharing across multiple data providers under the protection of privacy (Dedeoglu et al., 2020; Le Nguyen et al., 2020). Therefore, these characteristics enable blockchain to be an efficient and secure technical instrument for cross-border data exchanges (BSN, 2021). Some scholars (Chang et al., 2020; Liu and Li, 2020) have proposed the use of blockchain technology for cross-border trade. However, there are limited studies exploring the role of blockchain for cross-border data flow in the context of international trade. To the best of our knowledge, only Rahman et al. (2020) explored the use of permissioned blockchain for cross-border data sharing under a relaxed trust assumption. They proposed to build local data hubs on a cross-border data sharing platform and leverage the region's security gateway to record data on blockchain to mitigate data-centric misbehavior across countries. Although technically feasible, it is still in the proof-of-concept stage and is yet to be developed into a common platform to allow different blockchain systems to interact with each other across countries. Since April 2020, China's BSN has been emerging as the world's first cross-jurisdiction platform aiming to address the interoperability and regulatory compliance issues across national borders (BSN, 2021). In the following sections, this chapter investigates the BSN platform as a technical instrument in managing technical and regulatory risks pertaining to international trade and cross-border data flows and explores how the BSN infrastructure can interact with other solutions to enable cross-border data flows in the face of regulatory governance systems in force within different countries.

15.5 THE BLOCKCHAIN-BASED SERVICE NETWORK

BSN is a global initiative launched by China in April 2020 to create a technical ecosystem that can support cross-border data movements, utilizing blockchain data networks. The BSN Development Association has been formed by a government agency, two state-owned enterprises (SOEs), and a private sector software development firm as illustrated in Table 15.1. The State Information Centre of China (SIC), which leads the BSN initiative, is a Chinese government policy making think tank affiliated to the National Development and Reform Commission. China UnionPay is a Chinese financial services corporation headquartered in Shanghai. China Mobile is one of China's three main mobile carriers. Red Date Technology – a Hong Kong-based firm founded in 2014 – is the technical architect behind the BSN infrastructure.[3]

According to the *BSN User Manual*, the blockchain-based service infrastructure aims to provide developers and enterprises globally with a single 'one-stop-shop solution for blockchain and distributed ledger (DLT) applications', which can interoperate with each other within a single ecosystem (BSN Foundation, 2022). The ecosystem includes a range of resources for developers including gateway APIs, local SDKs, key certificates, etc. BSN

Table 15.1 BSN key development association members and their entities

	Government	Finance	Tech/software	Telecommunication
Association members	State Information Centre of China (SIC)[a]	China UnionPay Corporation[b] China Mobile Financial Technology Co., Ltd.[b]	Red Date Technology[c]	China Mobile Communications Corporation Design Institute Co., Ltd.[b] China Mobile Communications Corporation Government and Enterprise Service Company[b]

Source: BSN Development Association (2020a).

[a] government agency.
[b] Stated owned enterprise.
[c] Private enterprise.

---

[3] https://networking.report/c-suite-on-deck/interview-with-red-date-technology-yifan-he

has been variously framed as part of the Digital Silk Road infrastructure initiative (Sung, 2020) and has been described by Yaya Fanusie in testimony before the US–China Economic and Security Review Commission as providing infrastructure for new internet applications operating on shared data available to multiple parties in real time (Fanusie, 2021). In this sense, the BSN's design intent is to occupy the 'infrastructure or infrastructures' position, enabling the integration of cloud computing, 5G, AI, IoT, and big data together with financial technologies and other applications.

At the time of its launch in April 2020, BSN aimed to become a public and low-cost infrastructure for blockchain applications nationally and globally (Cao et al., 2020). According to BSN's proponents, businesses can save 80% of costs when deploying their blockchain-based business on the BSN platform (Stockton, 2020). In July 2020, the BSN platform offers a dual-pronged approach to serve both domestic and international networks. The China-based platform – known as BSN China – is governed by the BSN Development Association to provide services to domestic users, while the Singapore-based platform – known as BSN International – is governed by the BSN Foundation to offer blockchain application services to those outside China (Shen, 2021). The BSN International reflects the recognition that a single corporate entity responsible for a cross-border data exchange infrastructure/service is an impossible practice within China's rising regulatory landscape (Gkritsi, 2022; Gruin, 2021). BSN has been rapidly growing both nationally and globally. As of September 14, 2020, it has deployed 130 public city nodes in China and 15 public city nodes in countries, including Paris, Singapore, California, Sydney, and Tokyo.[4]

The BSN architecture comprises four key components: public city nodes (PCNs), blockchain frameworks, portals, and the network operations platform (Figure 15.1). According to the BSN Technical White Paper (BSN Development Association, 2020b), PCNs are a resource pool that allocates system resources from the cloud service or data center deployed to BSN. The PCNs mainly comprise these function modules, including blockchain frameworks, shared peer nodes, certificate authority (CA) management, authority chain, PCN gateway, and PCN manager systems. Blockchain frameworks support mainstream consortium blockchain frameworks (e.g. Fabric, FISCO, BCOS, CITA, Xuperchain, Wutong Chain, and Brochain) and public blockchain frameworks (e.g. Ethereum and EOS). BSN portals offer end users a blockchain-as-a-service platform. The BSN network operations platform offers supporting services, such as PCN management, billing and settlement management, CA management, and supervision management, to create a more flexible and stable service environment.

Real-life use cases for leveraging BSN to streamline cross-border data flows in accordance with local data privacy and security regulations are currently being developed between China and Singapore. According to

---

[4] https://hub.digitalasset.com/hubfs/20200907__BSN_DAML_PR_v7.pdf

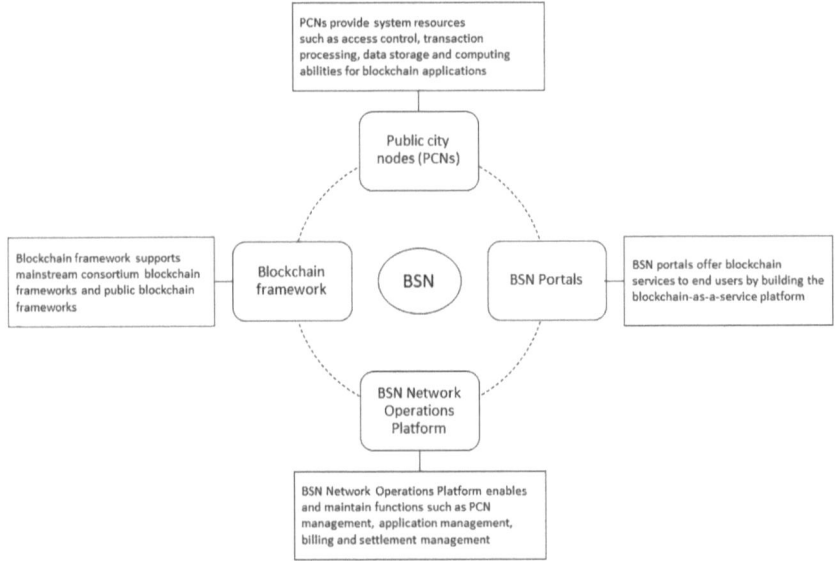

Figure 15.1 BSN architecture and key components (based on BSN Foundation 2022).

Gkritsi (2022), China's Shenzhen city and Singapore are developing the blockchain-based Transnational Trade Network (BBTN) built on the BSN network architecture, which allows companies to transfer cross-border trade data, while adhering to national regulatory compliance policies for protecting privacy and security. Although it is too early to say whether BBTN can develop into a multi-lateral network, it provides a potential framework that supports information security and enables compliance with jurisdictional data laws without disrupting cross-border trade data flows (Gkritsi, 2022).

The pilot use case of BBTN illustrates the potential of BSN to address the emergent demand for blockchain-based, cross-chain and cross-jurisdiction data exchange for trade. However, its trust-by-fiat model of permissioned mediation is at odds with the notion of trustworthiness expected of blockchain protocols (Cao et al., 2021b; Powell et al., 2021, 2022). To describe this using a legal analogy, participants to the BSN system are, first, required to request and obtain an authorization to interact with the system: this is known as 'permissioning' in the blockchain literature, a practice typically seen to be at odds with the unique value proposition of blockchains, which is to enable consensus in an adversarial setting in the absence of permissioning authority. Second, in order to avail BSN's data interchange services, registered participants are further required to delegate some or all of their credentials to the system, essentially creating what amounts to a power-of-attorney. Indeed, and unlike other permissioned chains which still have some sort of delegation accountability, BSN implements a very blunt form of delegation – at least in 'Trusted Mode', one of two offered modes of key management – where the system itself has the ability to cloak itself as any user and take action not just on the user's behalf but under their very identity, acting as an all-powerful 'Trusted Third Party' (TTP).

Technically speaking, BSN in 'Trusted Mode' vests in its federated city nodes the ability to generate and manage all of a user's keys on the participant's behalf. In particular, by allowing its central and federated authorities the ability to fully impersonate its users, BSN theoretically provides a turn-key mechanism for framing its users for actions they did not authorize, wholly undetected, not unlike traditional financial institutions. Both these features cast serious doubt on BSN's viability as a cross-jurisdiction mechanism for managing commerce since non-repudiation in the underlying blockchain presupposes that its participants have sole control of their private keys. While BSN aims to define a blockchain data interchange standard aiming to facilitate the flow of goods across jurisdictions by reducing frictions, technically speaking, BSN has not operated – and does not operate – any single blockchain protocol or network of its own. Rather, the BSN technical architecture can be described as a 'blockchain interchange', which seeks to create a common grammatical and linguistic environment to enable data tracked on different blockchains to be 'transported' from one to another.

We admit that BSN's technical architecture for cross-chain interoperability and data exchange reflects what is globally recognized as one of the most significant practical challenges to the wider adoption of blockchain (Robinson, 2021). However, from an information security standpoint, BSN in its current specification fails in key areas such as data confidentiality, sovereignty, and authenticity. Ultimately, BSN-mediated transactions will have to rely on the regulatory oversight and judicial redress of the choice-of law jurisdiction, rather than the mathematical assurances provided by the underlying blockchains. Confidentiality in the BSN framework is likely to remain an even thornier problem. The reliance on authority delegation through the blunt instrument of key escrow is undoubtedly an unfortunate design decision that goes against the basic principles of key management and information security. Incorruptible TTP authorities (in the information security sense) are unrealizable in the real world, and security cannot be enforced by fiat. It is indeed one of the goals of cryptography to design computer protocols which emulate the security of an ideal TTP without requiring one. In this technical sense, the use of a TTP, normally used as a notional tool to describe security, is being conflated with an ex-machina decree of the same – a category error. This raises new design issues due to the challenges for others to interact with the BSN infrastructure.

15.6 SQUARING THE CIRCLE THROUGH TRADE HUBS WITH THE AID OF CRYPTOGRAPHY

While the BSN, a permissioned blockchain 'framework' (not an actual chain), has rightly identified the issue of cross-jurisdictional data flow as a crucial friction point in cross-border trade, it fails to co-opt the trustworthiness benefits of permissionless blockchains. As such, the 'trust-by-fiat' BSN angle presents itself as a contrasting case study and strong motivator for seeking alternative hybrid – technology combined with policy – solutions that realize rather than impose verifiability of confidential data through technical means. Concretely, the unsolved challenge is to eschew reliance on centrally federated entities – BSN even asks users to cede complete control of their private keys – and instead to devise sound protocols for trustless cross-chain interoperations based on such cryptographic tools as encrypted commitments, digital signatures, and zero-knowledge proofs.

Real-world trade of material goods and services is finalized between multiple actors on the basis of data flows and information sharing (Cao et al., 2021a) – the original BSN motivation – and this involves further challenges. On one hand, with a focus on the safe exchange of confidential commercial data rather than of control of a cryptocurrency asset, the emphasis is not on the provision of an impregnable decentralized escrow system for the long-term safekeeping of staked cryptocurrency keys but on convincing commercial counterparts that a set of contractual conditions are met, without necessarily having to combine storage with secrecy. On the other hand, the required but heretofore missing requirement on this type of data exchange is an ability to prove that a statement is true, either to a designated counterpart or to any would-be verifier, while keeping under wraps what makes the statement true. For example, in a traditional finance scenario, a buyer or borrower would want to prove the availability of sufficient funds or income, without revealing their identity or the amount.

Paradoxical as it is, this is what the cryptographic notion of non-interactive zero-knowledge (NIZK) proof is able to achieve, with the ability to perform many different such proofs (Feige et al., 1999) with ease and agility across independent or even antagonistic blockchains. Using such computationally constrained tools as short smart contracts is the crux of the promising direction in blockchain-enabled trade, beyond BSN. Fortunately, these limitations are problems that cryptography is in principle able to resolve – if not in BSN, then in a successor or competitor – through the use of up-and-coming NIZK protocols. While they are not trivial, they are ideally suited to balance the dilemma between verifiability and confidentiality without having to trust any centrally controlling authority. We envision NIZK-based decentralized blockchain-based protocols as the answer to the BSN question – fulfilling the need for vibrant cross-jurisdictional trade without authoritative mediators – but daunting research and development challenges remain.

Seeking to square the circle of streamlining data exchange and meeting regulatory compliance across national boundaries, we propose a multi-agent cross-border trade consortium/framework in the form of "hub-to-hub" arrangements that are deployed on the BSN open permissioned blockchain network to reduce the intervention of authoritative mediators in the digital governance of cross-border trade. The design architecture for the BSN-based 'hub-to-hub' cross-border trade framework is illustrated in Figure 15.2. Our proposed architectural design is based on the BSN China and the BSN International to cater for the need of cross-border data exchanges in a multi-juridical environment. Trade Hub A is built on the BSN China with a China-based data center. Trade Hub B is built on the BSN International with a data center based outside China (e.g. Australia). The requirement of data localization can be met by setting up a data center in a specific trading country, which can enable data access for trade purposes via the connection by PCNs without violating the country-specific data regulatory agreement.

On the "hub-to-hub" design architecture, we seed advanced cryptographic techniques such as NIZK in both BSN ecosystems – BSN China and BSN International – to enable true cross-chain interchange of assets between two different blockchain networks. Trade members in each trade hub register to join the BSN network and enter the user portal to create a trade chain with their favorable blockchain framework, such as Etherum and Hyperledger Fabric. The established trade chain is connected to the data center, where paired smart contracts living on either side are called to watch each other and hold or release deposited assets in reciprocity, without

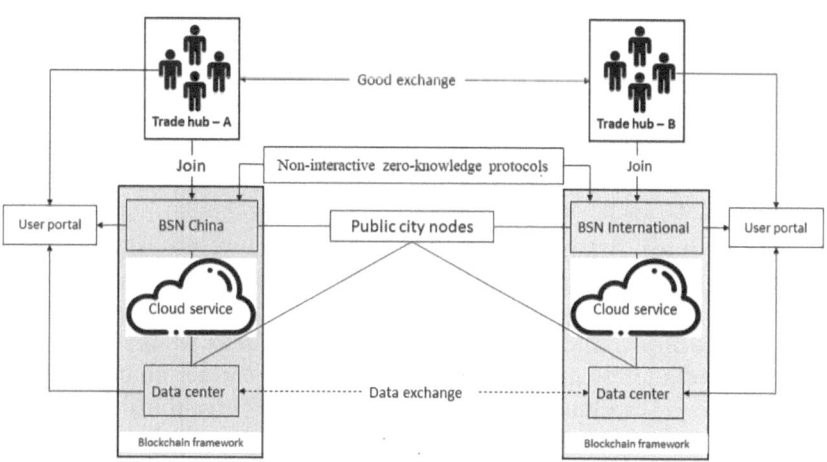

Figure 15.2 'Hub-to-hub' cross-border trade framework deployed on BSN.

requiring users to cede control of their asset to a TTP. The zero-knowledge proofs can ensure secure data-dependent cross-chain trade data, if only for a very specific application. The use of zero-knowledge proofs can 'prove that something is known without revealing the known information directly' (Hussey, 2020), therefore offering a solution to ensure data security and privacy, while making use of the value of the data for cross-border trade between businesses across countries. Such approaches are still in their infancy, being limited in both the scope of operation and the imperfect decentralization – which still requires non-permissioned but remunerated market makers to facilitate those trades and hold the deposited assets in escrow. Nonetheless, the direction for using technical instruments, particularly cryptographical tools, to resolve the conundrum of data sovereignty and privacy in data exchange is clear.

15.7 CONCLUSION

This chapter takes a technical perspective to explore BSN led by China, which seeks to enable cross-border trade with a hybrid approach in a jurisdictionally compliant environment. BSN defines a blockchain data interchange standard with the goal of facilitating the flow of goods across jurisdictions by reducing frictions on the flow of data. However, its approach is facilitated by a combination of permissioned access control with full key escrow, which is undoubtedly against the basic principles of distributed and decentralized key management and information security. To reduce the reliance on authority delegation in the data flow processes, this chapter proposed a techno-institutional proposition by introducing zero-knowledge proofs in a BSN-based 'hub-to-hub' cross-border trade framework. Despite being exploratory in nature, our chapter shows that technical instruments, particularly cryptographical tools, can help resolve the conundrum of data sovereignty and privacy in cross-border data exchange.

The exploratory yet critical nature of this chapter presents several areas for future research. First, deploying and evaluating the "hub-to-hub" cross-border trade framework on the BSN would be a practical research need when given that it helps provide evidence-based knowledge to unlock the potential of blockchain-based infrastructure – the BSN in this case – for cross-border data flow in the rising regulatory landscape. Second, incorporating NIZK proofs into the BSN design architecture is valuable from the perspective of addressing the limitation of the current design concept but could face several technical challenges. Third, despite technical feasibility, it is worth further investigating the acceptance of blockchain-based infrastructure – BSN in particular – across businesses and especially countries with different regulatory needs.

ACKNOWLEDGEMENT

This study was supported by funding from the Future Food Systems CRC Ltd., funded under the Australian Commonwealth Government CRC Program. The CRC Program supports industry-led collaborations between industry, researchers, and the community.

REFERENCES

Bauer, M., Erixon, F., Krol, M., Lee-Makiyama, H., & Verschelde, B. (2013). *The Economic Importance of Getting Data Protection Right: Protecting Privacy, Transmitting Data, Moving Commerce.* Retrieved February 10, 2022, from www.uschamber.com/assets/archived/images/documents/files/020508_EconomicImportance_Final_Revised_lr.pdf.

BSN. (2021). *Cross-Border Trade in the Age of China's Data Security Law.* Retrieved February 12, 2022, from https://medium.com/blockchain-thought-leadership/cross-border-trade-in-the-age-of-chinas-data-security-law-553b1115a785.

BSN Development Association. (2020a). *Blockchain-based Service Network (BSN) Introductory White Paper.* Retrieved February 10, 2022, from https://bsnbase.io/static/tmpFile/BSNIntroductionWhitepaper.pdf.

BSN Development Association. (2020b). *Blockchain-based Service Network Technical White Paper.* Retrieved March 5, 2022, from https://bsnbase.io/static/tmpFile/BSNTechnicalWhitePaper.pdf.

BSN Foundation. (2022). *Blockchain-based Service Network User Manual (Version 1.8.0).* Retrieved April 2, 2022, from https://bsnbase.io/static/tmpFile/bzsc/index.html?q=.

Bünz, B., Agrawal, S., Zamani, M., & Boneh, D. (2020). Zether: Towards privacy in a smart contract world. In *Financial Cryptography and Data Security* (pp. 423–443). Springer International Publishing. https://doi.org/10.1007/978-3-030-51280-4_23

Cao, S., Bryceson, K., & Hine, D. (2021). Collaborative risk management in decentralised multi-tier global food supply chains: an exploratory study. *The International Journal of Logistics Management, 32*(3), 1050–1067. https://doi.org/10.1108/IJLM-07-2020-0278

Cao, S., Deane, F., Foth, M., Powell, W., Robb, L., & Turner-Morris, C. (2020). China's Blockchain Services Network A new global emerging infrastructure. *ACS Information Age.* Retrieved February 20, 2022, from https://ia.acs.org.au/article/2020/blockchain-services-network.html.

Cao, S., Foth, M., Powell, W., & McQueenie, J. (2021a). What are the effects of short video storytelling in delivering blockchain-credentialed Australian beef products to China? *Foods, 10*(10). https://doi.org/10.3390/foods10102403.

Cao, S., Powell, W., Foth, M., Natanelov, V., Miller, T., & Dulleck, U. (2021b). Strengthening consumer trust in beef supply chain traceability with a blockchain-based human-machine reconcile mechanism. *Computers and Electronics in Agriculture, 180*, 105886. https://doi.org/10.1016/j.compag.2020.105886.

Casalini, F., & González, J. L. (2019). *Trade and Cross-Border Data Flows (OECD Trade Policy Papers). Organisation for Economic Co-Operation and Development (OECD)*. https://doi.org/10.1787/b2023a47–en.

Castro, D., & McQuinn, A. (2015). Cross-border data flows enable growth in all industries. *Information Technology and Innovation Foundation, 2*, 1–21.

Chang, Y., Iakovou, E., & Shi, W. (2020). Blockchain in global supply chains and cross border trade: a critical synthesis of the state-of-the-art, challenges and opportunities. *International Journal of Production Research, 58*(7), 2082–2099. https://doi.org/10.1080/00207543.2019.1651946.

Choi, J. H.-J., Foth, M., & Hearn, G. (2014). *Eat, Cook, Grow: Mixing Human-Computer Interactions with Human-Food Interactions*. MIT Press. https://doi.org/10.7551/mitpress/9371.001.0001.

Cory, N. (2017). *Cross-Border Data Flows: Where Are The Barriers, and What Do They Cost?* (pp. 1–42). Retrieved March 12, 2022, from https://itif.org/publications/2017/05/01/cross-border-data-flows-where-are-barriers-and-what-do-they.cost.

Cory, N., & Dascoli, L. (2021). *How Barriers to Cross-Border Data Flows Are Spreading Globally, What They Cost, and How to Address Them.* Information Technology and Innovation Foundation. Retrieved February 12, 2022, from https://itif.org/publications/2021/07/19/how-barriers-cross-border-data-flows-are-spreading-globally-what-they-cost.

Couture, S., & Toupin, S. (2019). What does the notion of "sovereignty" mean when referring to the digital? *New Media & Society, 21*(10), 2305–2322. https://doi.org/10.1177/1461444819865984.

Dedeoglu, V., Jurdak, R., Dorri, A., Lunardi, R. C., Michelin, R. A., Zorzo, A. F., & Kanhere, S. S. (2020). Blockchain technologies for IoT. In S. Kim & G. C. Deka (Eds.), *Advanced Applications of Blockchain Technology* (pp. 55–89). Springer Singapore. https://doi.org/10.1007/978-981-13-8775-3_3.

Fanusie, Y. J. (2021). *An Assessment of the CCP's Economic Ambitions, Plans, and Metrics of Success Panel IV: China's Pursuit of Leadership in Digital Currency.* Retrieved March 10, 2022, from https://www.uscc.gov/sites/default/files/2021-04/Yaya_Fanusie_Testimony.pdf.

Feige, U., Lapidot, D., & Shamir, A. (1999). Multiple noninteractive zero knowledge proofs under general assumptions. *SIAM Journal on Computing, 29*(1), 1–28. https://doi.org/10.1137/S0097539792230010.

Fernández-Caramès, T. M., & Fraga-Lamas, P. (2020). Towards post-quantum blockchain: A review on blockchain cryptography resistant to quantum computing attacks. *IEEE Access, 8*, 21091–21116. https://doi.org/10.1109/ACCESS.2020.2968985

Foth, M. (2017). The promise of blockchain technology for interaction design. In A. Soro, D. Vyas, J. Waycott, B. Ploderer, A. Morrison, & M. Brereton (Eds.), *Proceedings of the 29th Australian Computer-Human Interaction Conference (OzCHI 2017)* (pp. 513–517). Association for Computing Machinery. https://doi.org/10.1145/3152771.3156168.

Foth, M., Anastasiu, I., Mann, M., & Mitchell, P. (2021). From Automation to autonomy: technological sovereignty for better data care in smart cities. In B. T. Wang & C. M. Wang (Eds.), *Automating Cities: Design, Construction, Operation and Future Impact* (pp. 319–343). Springer Singapore. https://doi.org/10.1007/978-981-15-8670-5_13

Gallaher, M., & Cory, N. (2020). *4 Key Steps to Support Cross-Border Payments and Digital Trade Growth.* Retrieved March 5, 2022, from www.weforum.org/agenda/2020/06/action-on-cross-border-payments-will-support-digital-trade-growth/.

Gkritsi, E. (2022). *BSN's Red Date behind Shenzhen-Singapore Trade Blockchain Project.* Retrieved February 10, 2022, from www.coindesk.com/business/2022/01/18/bsns-red-date-behind-shenzhen-singapore-trade-blockchain-project/.

Gruin, J. (2021). The epistemic evolution of market authority: Big data, blockchain and China's neostatist challenge to neoliberalism. *Competition & Change,* 25(5), 580–604. https://doi.org/10.1177/1024529420965524

Hearn, G., Collie, N., Lyle, P., Choi, J., & Foth, M. (2014). Using communicative ecology theory to scope the emerging role of social media in the evolution of urban food systems. Futures, 62(Part B), 202–212. https://doi.org/10.1016/j.futures.2014.04.010.

Herian, R. (2020). Blockchain, GDPR, and fantasies of data sovereignty. *Law, Innovation and Technology,* 12(1), 156–174. https://doi.org/10.1080/17579 961.2020.1727094

Hussey, M. (2020, March 12). *What are Zero Knowledge Proofs?* Retrieved February 10, 2022, from https://decrypt.co/resources/zero-knowledge-proofs-explained-learn-guide.

Kay, L. (2019). *Data Flows in International Trade: What Are the Governance Options?* Retrieved February 15, 2022, from https://theodi.org/article/how-to-govern-data-flows-in-international-trade/.

Keogh, J. G., Rejeb, A., Khan, N., Dean, K., & Hand, K. J. (2020). Chapter 17 – Optimizing global food supply chains: The case for blockchain and GSI standards. In D. Detwiler (Ed.), *Building the Future of Food Safety Technology* (pp. 171–204). Academic Press. https://doi.org/10.1016/B978-0-12-818956-6.00017-8.

Kumar, A. (2020). *COVID-19: A Boon for the Digital Economy?* Retrieved February 20, 2022, from www.isb.edu/en/research-thought-leadership/research-centres-institutes/centre-for-learning-and-management-practice/management-rethink/covid-19--a-boon-for-the-digital-economy-html.

Laidlaw, E. (2021). *Privacy and Cybersecurity in Digital Trade: The Challenge of Cross Border Data Flows.* https://doi.org/10.2139/ssrn.3790936.

Lee-Makiyama, H. (2018). E-Commerce and digital trade. In J. Drake-Brockman & P. Messerlin (Eds.), *Potential Benefits of an Australia-EU Free Trade Agreement* (pp. 211–224). Retrieved February 12, 2022, from. https://digital.library.adelaide.edu.au/dspace/bitstream/2440/116956/2/hdl_116956.pdf#page=247.

Le Nguyen, B., Lydia, E. L., Elhoseny, M., Pustokhina, I., Pustokhin, D. A., Selim, M. M., Nguyen, G. N., & Shankar, K. (2020). Privacy preserving blockchain technique to achieve secure and reliable sharing of IoT data. *Computers, Materials & Continua,* 65(1), 87–107. https://doi.org/10.32604/cmc.2020.011599.

Liu, Z., & Li, Z. (2020). A blockchain-based framework of cross-border e-commerce supply chain. *International Journal of Information Management,* 52, 102059. https://doi.org/10.1016/j.ijinfomgt.2019.102059.

Mann, M., Mitchell, P., Foth, M., & Anastasiu, I. (2020). #BlockSidewalk to Barcelona: Technological sovereignty and the social license to operate smart cities. *Journal of the Association for Information Science and Technology*, *71*(9), 1103–1115. https://doi.org/10.1002/asi.24387.

Martinez, S. (2021). Could euro zone's GDPR serve as a multilateral regulatory framework on cross-border data flows? *Business and Public Administration Studies*, *15*(1), 24–26. https://www.bpastudies.org/index.php/bpastudies/article/view/250.

Meese, J., Jagasia, P., & Arvanitakis, J. (2019). Citizen or consumer? Contrasting Australia and Europe's data protection policies. *Internet Policy Review*, *8*(2). https://policyreview.info/articles/analysis/citizen-or-consumer-contrasting-australia-and-europes-data-protection-policies.

Meltzer, J. P. (2015a). *A New Digital Trade Agenda*. Retrieved February 08, 2022, from http://e15initiative.org/wp-content/uploads/2015/07/E15-Digital-Economy-Meltzer-Overview-FINAL.pdf.

Meltzer, J. P. (2015b). The internet, cross-border data flows and international trade. *Asia & the Pacific Policy Studies*, *2*(1), 90–102. https://doi.org/10.1002/app5.60.

Meltzer, J. P., & Lovelock, P. (2018). *Regulating for a Digital Economy: Understanding the Importance of Cross-Border Data Flows in Asia*. Brookings. Retrieved February 10, 2022, from https://www.brookings.edu/research/regulating-for-a-digital-economy-understanding-the-importance-of-cross-border-data-flows-in-asia/.

Mitchell, A. D., & Mishra, N. (2018). Data at the docks: Modernizing international trade law for the digital economy. *Vanderbilt Journal of Entertainment and Technology Law*, *20*(4), 1073–1134. https://scholarship.law.vanderbilt.edu/jetlaw/vol20/iss4/3/.

Mitchell, A. D., & Mishra, N. (2019). Regulating cross-border data flows in a data-driven world: How WTO law can contribute. *Journal of International Economic Law*, *22*(3), 389–416. https://doi.org/10.1093/jiel/jgz016.

Morozov, E. (2014). *To Save Everything, Click Here: The Folly of Technological Solutionism*. PublicAffairs.

National Board of Trade. (2014). *No Transfer, No Trade – the Importance of Cross-Border Data Transfers for Companies Based in Sweden*. Retrieved February 20, 2022, from https://unctad.org/system/files/non-official-document/dtl_ict4d2016c01_Kommerskollegium_en.pdf.

Nguyen, D., & Paczos, M. (2020). Measuring the economic value of data and cross-border data flows: A business perspective. *OECD Digital Economy Papers*, *297*, 1–47. https://doi.org/10.1787/6345995e-en

Pepper, R., Garrity, J., & LaSalle, C. (2016). Cross-border data flows, digital innovation, and economic growth. *The Global Information Technology Report*, 39–47. https://www3.weforum.org/docs/GITR2016/WEF_GITR_Chapter1.2_2016.pdf.

Powell, W., Cao, S., Miller, T., Foth, M., Boyen, X., Earsman, B., del Valle, S., & Turner-Morris, C. (2021). From premise to practice of social consensus: How to agree on common knowledge in blockchain-enabled supply chains. *Computer Networks*, *200*, 108536. https://doi.org/10.1016/j.comnet.2021.108536.

Powell, W., Foth, M., Cao, S., & Natanelov, V. (2022). Garbage in garbage out: The precarious link between IoT and blockchain in food supply chains. *Journal of Industrial Information Integration*, *25*, 100261. https://doi.org/10.1016/j.jii.2021.100261.

Rahman, M. S., Al Omar, A., Bhuiyan, M. Z. A., Basu, A., Kiyomoto, S., & Wang, G. (2020). Accountable cross-border data sharing using blockchain under relaxed trust assumption. *IEEE Transactions on Engineering Management*, *67*(4), 1476–1486. https://doi.org/10.1109/TEM.2019.2960829.

Robinson, P. (2021). Survey of crosschain communications protocols. *Computer Networks*, *200*, 108488. https://doi.org/10.1016/j.comnet.2021.108488.

Roxas, H. (2021). *Traceability and Trusted Data for Supply Chains*. Retrieved February 15, 2022, from www.finextra.com/blogposting/20467/traceability-and-trusted-data-for-supply-chains.

Shen, T. (2021). *China's Blockchain Infrastructure Project Goes Live in Hong Kong, Macau*. Retrieved February 22, 2022, from https://forkast.news/chinas-blockchain-infrastructure-project-goes-live-in-hong-kong-macau/.

Snelson, S., & Cilauro, F. (2021). *Unlocking the Economic and Social Value of Private and Third Sector Data*. Retrieved April 02, 2022, from www.frontier-economics.com/uk/en/news-and-articles/news/news-article-i8238-unlocking-the-economic-and-social-value-of-private-and-third-sector-data/.

Steger, M. B., & Roy, R. K. (2010). *Neoliberalism: A Very Short Introduction*. Oxford University Press.

Stockton, N. (2020). China takes blockchain national: The state-sponsored platform will launch in 100 cities. *IEEE Spectrum*, *57*(4), 11–12. https://doi.org/10.1109/MSPEC.2020.9055903.

Sung, M. (2020). *China's National Blockchain Will Change the World*. Retrieved February 10, 2022, from www.coindesk.com/policy/2020/04/24/chinas-national-blockchain-will-change-the-world/.

Thumfart, J. (2022). The norm development of digital sovereignty between China, Russia, the EU and the US: From the late 1990s to the COVID crisis 2020/21 as catalytic event. In D. Hallinan, R. Leenes, & P. De Hert (Eds.), *Data Protection and Privacy, Volume 14: Enforcing Rights in a Changing World*. Bloomsbury Academic. https://doi.org/10.5040/9781509954544.ch-001.

UNCTAD. (2021). *Digital Economy Report 2021 – Cross-Border data Flows and Development: For Whom the Data Flow*. Retrieved February 15, 2022, from https://unctad.org/system/files/official-document/der2021_en.pdf.

United States International Trade Commission. (2014). *Digital Trade in the U.S. and Global Economies, Part 2*. Retrieved February 12, 2022, from www.usitc.gov/publications/332/pub4485.pdf.

Valeonti, F., Bikakis, A., Terras, M., Speed, C., Hudson-Smith, A., & Chalkias, K. (2021). Crypto collectibles, museum funding and OpenGLAM: Challenges, opportunities and the potential of non-fungible tokens (NFTs). *Applied Sciences*, *11*(21), 9931. https://doi.org/10.3390/app11219931.

Index